**PAUL R. JONES**
*University Of New Hampshire*
*Department Of Chemistry*
*Durham, New Hampshire 03824*

properties
and processing
of polymers
for engineers

PRENTICE-HALL, INC.
Englewood Cliffs, NJ
07632

 SOCIETY OF PLASTICS
ENGINEERS, INC.

# properties
# and processing
# of polymers
# for engineers

G.R. MOORE
D.E. KLINE

*Library of Congress Cataloging in Publication Data*

MOORE, G. R. (GREGORY R.)   date
    Properties and processing of polymers for engineers.

    Includes bibliographical references and index.
    1. Polymers and polymerization.   I. Kline, D. E.
(Donald Edgar), 1928-      .   II. Title.
QD381.M64 1984       668.9       83-9604
ISBN 0-13-731125-7

*Chem*
*QD*
*381*
*.M64*
*1984*

Editorial/production supervision
    and interior design: **Karen Skrable**
Manufacturing buyer: **Anthony Caruso**

Printed in the United States of America

10   9   8   7   6   5   4   3   2   1

ISBN 0-13-731125-7

**PRENTICE-HALL INTERNATIONAL, INC.,** *London*
**PRENTICE-HALL OF AUSTRALIA PTY. LIMITED,** *Sydney*
**EDITORA PRENTICE-HALL DO BRASIL, LTDA.,** *Rio de Janeiro*
**PRENTICE-HALL CANADA INC.,** *Toronto*
**PRENTICE-HALL OF INDIA PRIVATE LIMITED,** *New Delhi*
**PRENTICE-HALL OF JAPAN, INC.,** *Tokyo*
**PRENTICE-HALL OF SOUTHEAST ASIA PTE. LTD.,** *Singapore*
**WHITEHALL BOOKS LIMITED,** *Wellington, New Zealand*

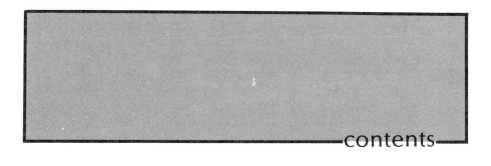

# contents

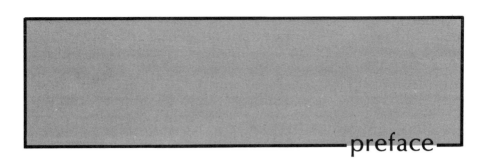

preface

In recent years the need for basic information on the manufacture, design, and use of polymers has become increasingly evident. The annual consumption of these materials is now greater than that of steel and, by one estimate[1], is projected to exceed that of all metals by 1990. Contact with polymers in engineering occupations is therefore nearly inevitable. The fact that these materials exhibit vastly different properties from more traditional engineering materials underscores the need for a better understanding of their behavior.

The material in this book was developed as a single-credit course at the Pennsylvania State University. Its purpose is to introduce mechanical and industrial engineering students to polymers. It can also be adapted for students in other polymer-related disciplines such as wood technology and biomedical engineering, with the addition of appropriate examples. The principles in this text are fundamental to polymers and are therefore as applicable to the design of a prosthetic device as they are to a wood adhesive or to an automobile component. The examples in this text are slanted toward mechanical and industrial engineers, in keeping with the original intent of the course.

In teaching we have found it useful to require that the students learn the structures of some polymers. This makes the lectures easier to present and also helps the students to gradually associate specific properties (e.g., melt and glass temperatures, hygroscopicity, crystallinity) with these structures. In this way the students become familiar with the properties and behavior of the commercially

---

[1]"More Polymer Than Metal Will Be Consumed by 1990." *Plastics Engineering* (1980):27.

important polymers; the repeating units serve as a memory aid instead of being an exercise in retention.

The course and this text are based on the philosophy that one can learn a great deal by asking the proper questions. In past years we presented many of the questions for the students in the form of problem sets and required that the students answer them in anticipation of a quiz the following period. We have included these problem sets at the end of each chapter. Although some of the questions seem simplistic, experience has taught us that this is not the case for those viewing the material for the first time.

Some of the questions require information that is not explicitly contained in this text. This is done deliberately to encourage students to explore polymer-related references. In particular we suggest that students become aware of (1) *Modern Plastics Encyclopedia*, (2) *ASTM Standards*, (3) *Chemical Marketing Reporter*, and (4) a few detailed polymer textbooks. In cases where these references are helpful, notes to that effect are included.

We have also found that an excessive amount of reading per assignment tends to discourage interest and progress at an introductory level. For this reason we have tried to keep the chapters short. This has necessarily resulted in the omission of much detail. Wherever possible, however, we cite more advanced references that can provide the interested student with this additional material.

G.R. MOORE
D.E. KLINE

# acknowledgments

We would like to thank our colleagues, Drs. Ian R. Harrison and Michael M. Coleman, for their comments concerning the manuscript. We would also like to thank the many students in Polymer Science 210 for their willingness to work with the rough draft and to provide numerous constructive comments. Finally, we are very grateful to Mrs. Gail Harvey, who typed the original draft of the manuscript.

# section i

# introduction

# the importance of polymers

## chapter 1

Without polymers the world as we know it would be different, if it could even exist at all. Living things contain polymers, including the DNA that codes the life processes in our bodies and the cellulose and other components that make up most plant life. Wood, bark, leaves, skin, feathers, bone, and silk are just some of the naturally occurring polymers associated with life today.

To this large volume of polymers, which has been regenerated for millions of years, we have added many new polymers. Initially these additions included a wide variety of modified forms of natural polymers such as leather, paper, natural rubber, rayon, and the cellulose acetates and nitrates. Within the past 100 years, however, the trend has shifted somewhat from the direct use and modification of natural polymers to the production of totally synthetic polymers from materials of low molecular weight such as oil and the pyrolysis products of coal and wood. These are the polyethylenes, nylons, polyesters, etc., which have become an integral part of our lives. These synthetic polymers, together with naturally occurring forms, represent a substantial portion of the materials production (Table 1.1) in this country.

Synthetic polymers have a wide range of applications, and it would be difficult for us to do without them in many of these. Our lifestyles would be changed significantly, for example, without the use of elastomers for automobile tires. Similarly, the polymers used in the insulation of electrical and electronic components would be difficult to replace. In addition to these critical applications, however, synthetic polymers have found their way into almost every other aspect of our daily lives. These range from the packaging used to protect food to

1

**TABLE 1.1  Examples of Annual Materials Production in the United States (late 1970s)**

| Material | Production | Approximate Volume |
|---|---|---|
| Metals | | |
| Raw steel | $1.2 \times 10^{11}$ kg | $1.6 \times 10^7$ m$^3$ |
| Aluminum | $4.4 \times 10^{10}$ kg | $1.6 \times 10^7$ m$^3$ |
| Copper | $1.4 \times 10^9$ kg | $1.6 \times 10^5$ m$^3$ |
| Lead | $5.3 \times 10^8$ kg | $4.2 \times 10^4$ m$^3$ |
| Polymers | | |
| Softwood lumber and plywood | $5.6 \times 10^{10}$ kg | $1.0 \times 10^8$ m$^3$ |
| Synthetics (excluding elastomers and textile fibers) | $1.5 \times 10^{10}$ kg | $1.5 \times 10^7$ m$^3$ |
| *Elastomers (natural and synthetic) | $3.5 \times 10^9$ kg | $3.5 \times 10^6$ m$^3$ |
| Fibers (synthetic) | $4.7 \times 10^9$ kg | $4.7 \times 10^6$ m$^3$ |
| Fibers (cotton and wool) | $2.1 \times 10^9$ kg | $2.1 \times 10^6$ m$^3$ |
| Minerals | | |
| Portland cement | $7.2 \times 10^{10}$ kg | $2.3 \times 10^7$ m$^3$ |
| Sand and gravel | $8.2 \times 10^{11}$ kg | $3.4 \times 10^8$ m$^3$ |

*Combined U.S. and Canadian consumption, rather than production. (*Adapted from* [1], [2], *and* [3].)

the materials of which our houses, clothing, furniture, cars, and appliances are made.

Additional uses for polymers are constantly being found. Many of these involve materials substitution problems such as the use of lighter-weight polymeric components to replace steel and other materials in automobiles, and the replacement of glass beverage containers with polyester ones. Other applications, on the other hand, involve completely new products. Some examples of these products include dust-control agents for pigments and other finely divided industrial materials, artificial hip and finger joints for people crippled with arthritis, and polymeric additives used to strengthen concrete in bridges and other structures.

The success of these and future applications depends to a large extent on scientists' and engineers' ability to understand and utilize these materials properly. This text is intended to promote this understanding, both by serving as an introduction to these materials and by providing encouragement for students to pursue selected topics of interest in detail.

## REFERENCES

[1] *The World Almanac and Book of Facts*, Newspaper Enterprise Assoc., New York (1979).

[2] Seymour, R. B., and C. E. Carraher, Jr., *Polymer Chemistry*, Dekker, New York (1981).

[3] Hoyle, R. J., *Wood Technology in the Design of Structures*, 4th ed. Mountain Press Publ. Co., Missoula, Montana (1978).

Since most polymers are organic molecules, some familiarity with the principles of organic chemistry is necessary in order to understand them. The material presented here should be reviewed before proceeding to other chapters.

## ELEMENTS

The element of major importance in organic polymers is carbon, whose symbol is C. It has a valence (number of bonding electrons) of 4, and it often bonds to other atoms (including carbon) in a pyramidal fashion with bond angles of 109.5°.

109.5°

Bond Length = $1.54 \times 10^{-10}$ m
Bond Energy* = 349 kJ/mol

In addition to single bonds, carbon will also form double bonds (which involve the sharing of four electrons instead of two) with an adjacent carbon atom:

Bond Length = $1.32 \times 10^{-10}$ m
Bond Energy – 607 kJ/mol

*Bond energy refers to the strength of the bond. It is expressed as the energy (Joules) required to break $6.02 \times 10^{23}$ (1 mole) of these bonds.

and triple bonds (which involve the sharing of six electrons) with an adjacent carbon atom:

$$-C \equiv C- \qquad \begin{array}{l} \text{Bond Length} = 1.20 \times 10^{-10} \text{ m} \\ \text{Bond Energy} = 828 \text{ kJ/mol} \end{array}$$

Double and triple bonds differ from single bonds not only in their higher strengths and shorter bond lengths but also in rotatability. A single bond will permit rotation, whereas double and triple bonds are rigid and will not.

$$-\overset{\mid}{C}\left(\overset{\mid}{C}-\right. \qquad\qquad\qquad >C\!\!\times\!\!C< -C\!\!\times\!\!C-$$

|  single bond  |  double and triple bonds  |
| :---: | :---: |
| (rotation is possible) | (rotation is *not* possible) |

As will be seen later, this ability (or inability) to rotate is important in determining the flexibility of a polymer molecule. This flexibility in turn has a marked effect on the thermal and mechanical properties of the bulk material.

A polymer molecule usually consists of carbon and one or more other elements. These other elements are:

| Hydrogen | $H-$ |
| :--- | :--- |
| Oxygen | $-O-$ |
| Nitrogen | $\diagup N \diagdown$ |
| Chlorine | $Cl-$ |
| Fluorine | $F-$ |
| | |
| Sulfur | $-S-$ and $-\overset{\|}{\underset{\|}{S}}-$ |

Occasionally silicon ($-\overset{\mid}{\underset{\mid}{Si}}-$) or phosphorus ($-\overset{\mid}{\underset{\mid}{P}}{=}$) may be present.

It is important to keep some characteristics of these elements (such as atomic weights and common valences) in mind. Since it is easy to forget them if they are only used occasionally, a summary is included in Appendix A. Such information is also readily obtained from a periodic table. One must keep in mind that these elements are real and have sizes that are functions of their positions in the periodic table. For example, chlorine is a large atom compared to fluorine or hydrogen because it has a larger nucleus and a greater number of electron shells. This implies that a chlorine atom attached to a polymer molecule will tend to interfere with the movement of other molecules much more than a smaller atom would. As will be seen later, this can affect many properties, including the amount of crystallinity that exists, the modulus of the polymer, and the softening temperature.

## CHEMICAL BONDING

There are two types of bonds in polymers:

1. Primary bonds, which hold together the individual atoms that make up a polymer molecule (chain), and
2. Secondary bonds, which provide the attraction between molecules.

Primary bonds are covalent (electron sharing). Secondary bonds are usually one or more of the following:

1. Van der Waals forces, which occur between all molecules. They are the weakest (2.1–8.4 kJ/mol).
2. Dipole interactions, which occur when polar molecules are involved. They can be much stronger than the van der Waals forces (6.3–12.6 kJ/mol).
3. Hydrogen bonds (which are a special case of dipole interactions) involving hydrogen and the strongly electronegative elements: oxygen, nitrogen, or fluorine. They are the strongest secondary bonds (12.6–29.3 kJ/mol).

It is not necessary that one understand the origins of these secondary bonds for the purposes of this text. However, it is important to recognize which atoms or groups of atoms contribute to their occurrence. Figure 2.1 is a rough guide for this.

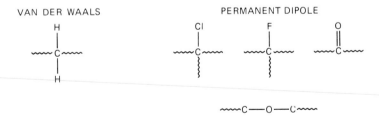

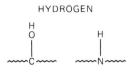

**Figure 2.1** Some Molecular Groups Associated with Various Types of Secondary Bonding in Polymers.

## ORGANIC CHEMISTRY NOMENCLATURE

The names of organic compounds sometimes seem complicated. They need not be overwhelming, however. There are a number of terms that are frequently used and with which the student should become familiar. A brief summary of these terms is given in Appendix B. As these terms occur in the reading, the student should refer to this appendix for a brief definition. Most of the basic chemistry-related information needed to understand the material in this text is contained in this chapter and in Appendices A and B.

A word of caution is appropriate at this point. Many organic compounds can be extremely dangerous. Since they are not fully oxidized, they often present fire hazards. In addition, many are toxic and some are potential carcinogens. Thus, before working with any of these chemicals, the student should become familiar with the appropriate safety procedures. For information on flammability and explosivity, a good reference is the National Fire Protection Association's *Fire Protection Guide on Hazardous Materials* [1]. For information on toxicity, a good reference is *Dangerous Properties of Industrial Materials* by N. Irving Sax [2].

## PROBLEM SET

1. What are six common atoms found in organic polymers? Locate each on a periodic table. What are the valences and atomic weights of each? Which is the largest? Which is the smallest?

2. Distinguish between primary and secondary bonding in polymers.

3. What is meant by thermal degradation? How are primary and secondary bonds involved in this phenomenon?

**Answer:**

Thermal degradation is the destruction and alteration of primary (covalent) bonds in molecules. This change is caused by excessive temperature, as opposed to ultraviolet light or chemical agents which cause other types of degradation. In this process the chemical structure of the molecules is permanently changed. Thermal degradation does not directly involve the nature of the secondary bonds. This is discussed in greater detail in Chapter 9.

4. What is meant by dissolving? How are primary and secondary bonds involved?

**Answer:**

Dissolving is the dispersion of the molecules of one substance (called the solute) uniformly throughout the molecules of a second substance (called the solvent). Usually, the solute is a solid and the solvent is a liquid. This process involves the disruption of the secondary bonds in the solid and the replacement of these bonds with solvent–solute bonds. This does not change the primary chemical structure of the molecules. Primary bonds are also involved in dissolving, but in a rather subtle manner (see Chapter 11).

5. What type(s) of intermolecular bonding (van der Waals, dipole, or hydrogen bonds) predominate in the following compounds?

(a) Water               (e) Methanol
(b) Gasoline           (f) Acetone
(c) Benzene           (g) Methane
(d) Vinyl chloride      (h) Nitrogen

Why?

6. Methyl methacrylate, styrene, and vinyl chloride are common organic compounds used to make polymers. Assume that you had to work with these materials on a daily basis. How do they compare as regards:

(a) Their fire hazard?
(b) Their tolerable exposure levels (skin, inhalation, ingestion)?

7. Show that the tetrahedral bond angle for the carbon atom is 109.5°.

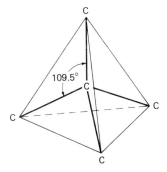

8. Why is a hydrogen bond stronger than a permanent dipole?

**Answer:**

A hydrogen bond is a special permanent dipole. Hydrogen atoms consist of only one proton and one electron. Thus, when a hydrogen atom is covalently bonded to a highly electronegative atom (oxygen, nitrogen, or fluorine), its electron is pulled almost completely away from its proton. This exposes the positive proton, which can then interact rather strongly with electronegative atoms on neighboring molecules. In materials containing permanent dipoles (which are not hydrogen bonds), electron clouds shield the positive nuclei somewhat. This means that the secondary permanent dipole bonds which form will be weaker than those in hydrogen-bonded materials.

9. How do metals, ceramics, and molecular solids (such as sugar) differ with regard to their chemical bonding?

10. Can we produce a new variation (i.e., an isomer) of a compound or polymer by rotation about a bond? Why?

**Answer:**

Molecules that can be converted from one type to another solely by rotation about single bonds are conformational isomers. These are currently not of great importance in synthetic polymers, although they are important in biological systems (blood proteins, etc.). For the purposes of this text, rotation about single bonds does not change the molecule or its properties.

## REFERENCES

[1] National Fire Protection Association, *Fire Protection Guide on Hazardous Materials*, 7th ed., Boston (1978).

[2] Sax, N. I., *Dangerous Properties of Industrial Materials*, 5th ed., Van Nostrand Rheinhold, New York (1979).

# characterization of polymers

# the chemical structure of polymers

## chapter 3

## POLYMERS

The word *polymer* is derived from the Greek *polys*, meaning numerous, and *meros*, meaning part. Saunders [1] defines a polymer as a large molecule made of one or more repeating units linked together by covalent bonds. The term *macromolecule* (from the Greek *macros*, meaning long) is often used as a synonym, although it is not strictly correct, since it only conveys the idea of a large molecule and does not take into account the concept of a repeating unit.

Two points need clarification: What is a repeating unit? And how large is "large"? A repeating unit is simply a group of atoms linked together covalently in a particular spatial arrangement. For example, polyvinyl chloride (PVC, vinyl) has the following repeating unit:

$$\left[\begin{array}{cc} H & Cl \\ | & | \\ -C-C- \\ | & | \\ H & H \end{array}\right]$$

and a molecule of PVC is a series of these units covalently bonded together:

$$\sim\sim C - C - C - C - C - C - C - C \sim\sim$$

To indicate the molecule, it is usually convenient to use the shorthand notation

$$\left[\begin{matrix} \overset{\displaystyle H}{\underset{\displaystyle H}{\rule{0pt}{1em}}} & \overset{\displaystyle Cl}{\underset{\displaystyle H}{\rule{0pt}{1em}}} \\ -C-C- \end{matrix}\right]_n$$

where the subscript $n$ indicates the number of times the repeating unit occurs within the polymer. This subscript is termed the degree of polymerization (d.p.).

The length of a polymer molecule is directly related to $n$, and polymers are often regarded as molecules with $n > 100$. This is an arbitrary number, however, and commercial polymers generally contain many more than 100 units. For example, a typical molecular weight for a PVC molecule is 50,000 g/mol. This means that $6.02 \times 10^{23}$ of these molecules weigh 50,000 grams, or roughly 110 pounds. To calculate the corresponding number of repeating units, one uses the following relationship:

Molecular Weight of Polymer

$$= \text{d.p.} \times (\text{Molecular Weight of Repeating Unit})$$

The repeating unit of PVC consists of 2 carbon atoms, 3 hydrogen atoms, and 1 chlorine atom. Its molecular weight is therefore

$$2 \times 12 + 3 \times 1 + 35.5 = 62.5$$

and the degree of polymerization is $50,000/62.5 = 800$.

It is generally accepted that some of the largest synthetic linear polymer molecules are those of Teflon* (polytetrafluoroethylene), whose repeating unit is

$$\left[\begin{matrix} \overset{\displaystyle F}{\underset{\displaystyle F}{\rule{0pt}{1em}}} & \overset{\displaystyle F}{\underset{\displaystyle F}{\rule{0pt}{1em}}} \\ -C-C- \end{matrix}\right]_n$$

Molecular weights of $3 \times 10^6$ g/mol are not unusual for this polymer. Using the method outlined above, one can readily see that this corresponds to a degree of polymerization of 30,000.

While this number is rather large, its significance is more readily seen in graphical terms. For instance, if this dot, $\frac{1}{8}$ inch in diameter, were the size of an iron atom in a piece of steel ($\sim 2.5 \times 10^{-10}$ m)

then on the same size scale the Teflon molecule would be approximately 127 meters (or about 400 feet) long! The fact that polymer molecules are so large compared to other atomic particles is a major factor in their unusual behavior.**

*Teflon is a du Pont product.

**Any reader who is uncertain as to the distinction among metals, ceramics, and molecular solids should consult a basic textbook on chemistry. It is important to recognize these basic differences.

The concept of a very large molecule whose structure is repetitive probably seems reasonable at this point. This certainly was not the case in the late nineteenth and early twentieth centuries. As late as the 1920s it was generally thought that polymers were compounds of low molecular weight and that their unusual behavior was due to impurities forcing them to aggregate in a colloidal form. This view was questioned in 1920 by Staudinger, who proposed long chain structures as an alternative. The history behind the concept and development of polymers is an interesting one, and students are encouraged to review it [2].

## MONOMERS

Another term closely associated with polymers is *monomer.* A monomer is the molecular compound that is used to produce the polymer. In the case of polyvinyl chloride (PVC), for example, the monomer is vinyl chloride*, a colorless gas with a boiling point of $-14°C$ at atmospheric pressure. It has the following structure:

$$\begin{array}{c} H \\ \diagdown \\ H \end{array} C = C \begin{array}{c} Cl \\ \diagup \\ H \end{array}$$

The polymer PVC is formed from this monomer via the process of polymerization, which can be illustrated schematically as

$$ n \left( \begin{array}{c} H \\ \diagdown \\ H \end{array} C = C \begin{array}{c} Cl \\ \diagup \\ H \end{array} \right) \longrightarrow \left[ \begin{array}{cc} H & Cl \\ | & | \\ C - C \\ | & | \\ H & H \end{array} \right]_n $$

The actual process will be explained in detail in a later chapter. At this point, however, it is important to recognize that the monomer and the repeating unit are not the same because of changes that occur during the polymerization process.

A basic property of all monomers is functionality. This is the number of sites at which the monomer molecule can link with other monomer molecules to form a long chain. For example, vinyl chloride has a functionality of 2 in the polymerization of PVC. This means that it will link chemically in two places:

$$ \longrightarrow \begin{array}{c} H \\ \diagdown \\ H \end{array} C = C \begin{array}{c} Cl \\ \diagup \\ H \end{array} \longleftarrow $$

and the result will usually be a linear chain. This is also the case with polymers such as polytetrafluoroethylene, polymethyl methacrylate, and polypropylene whose monomers are also based on the $\diagup\!C\!=\!C\!\diagdown$ structure.

---

*In recent years vinyl chloride has become recognized as a carcinogenic material. Consequently its permissible exposure levels are extremely low. Despite the high cost of adhering to these strict levels, PVC continues to be one of the highest-volume synthetic polymers produced in the United States.

In cases where the functionality is greater than 2, the resulting polymer will be branched; that is, side chains will grow from the main chain. If a sufficient number of these higher-functionality units exist in the chain, then all of the main chains become linked together through these side chains. These are *crosslinked polymers*. As an example, consider the use of methyl methacrylate (MMA) as an impregnant for a porous material such as concrete. Under normal circumstances MMA polymerizes as a linear chain via the following process:

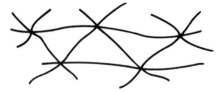

When it is used as an impregnant, a low percentage of TMPTMA (trimethylol-propane trimethacrylate) is often added in order to crosslink the otherwise linear molecules. As will be seen later, this imparts certain desirable properties to the system in question. The structure of TMPTMA is

Since the $\overset{}{\underset{}{>}}C{=}C\overset{}{\underset{}{<}}$ groups are the reactive ones, TMPTMA has a functionality of 6 in the reaction with MMA. In graphical terms the completed system will have a structure like the following with the TMPTMA added:

as opposed to the following in the uncrosslinked system:

Crosslinking and its consequences will be examined shortly in greater detail. Its significance at this point is the degree to which it is related to the concept of functionality.

## COPOLYMERS

Most of the polymers described thus far are those derived from essentially one type of monomer. That is, the resulting polymer contains only one type of repeating unit (except the MMA/TMPTMA example). These polymers are known as homopolymers. It is possible, however (as in the MMA/TMPTMA example), to have polymers that contain more than one type of repeating unit in the polymer chain. These are called *copolymers*. A moment's reflection upon the possibilities of joining just two types of repeating units will readily reveal four options. If A and B designate the two types of repeating units, they could, for example, occur randomly along the chain, as in

AABABABBABAA

They could also alternate:

ABABABABABAB

They could occur in blocks of various lengths, as in

AAAAABBBBBAAAAABBBB

And finally, they could branch:

```
AAAAAAAAAAAAAA
B             B
B             B
B             B
B             B
B             B
B             B
```

These are *random*, *alternating*, *block*, and *graft* copolymers, respectively. As will be seen later, copolymerization extends the range of properties available to the engineer because of the additional structural variables it imparts to the molecules.

It is also possible to have more than two types of repeating units in the chain. A polymer in which three occur is called a *terpolymer*. Relatively few of these exist, although one of increasing importance is the EPDM (ethylene propylene diene terpolymer) elastomer (Chapter 14).

## TYPES OF POLYMERS

It has been estimated that there are approximately 30,000 patented polymers in the United States. It seems fortunate for all of us, however, that only a few have any significant commercial volume. In 1979, for example, U.S. production of

TABLE 3.1  **Sales of Synthetic Polymers in the United States (1979)**

| Polymer | Megagrams Sold |
|---|---|
| Low-density polyethylene | $3.18 \times 10^6$ |
| Polyvinyl chloride and its copolymers | $2.53 \times 10^6$ |
| High-density polyethylene | $1.98 \times 10^6$ |
| Polystyrene | $1.66 \times 10^6$ |
| Polypropylene and its copolymers | $1.60 \times 10^6$ |
| Polyesters (thermoplastic and thermoset) | $0.88 \times 10^6$ |
| Polyurethane foams | $0.79 \times 10^6$ |
| Phenolics | $0.64 \times 10^6$ |
| Urea and melamine | $0.53 \times 10^6$ |
| ABS | $0.54 \times 10^6$ |
| Acrylic (includes PMMA) | $0.24 \times 10^6$ |
| Nylons | $0.13 \times 10^6$ |
| Polyacetal (polyformaldehyde and others) | $0.04 \times 10^6$ |
| Polycarbonate | $0.10 \times 10^6$ |
| Epoxy | $0.14 \times 10^6$ |
| Others (includes polypropylene oxide and polysulfone) | $1.00 \times 10^6$ |

(*Adapted from* [3].)

synthetic polymers was reported as shown in Table 3.1. For future reference the repeating units of these and several other polymers are given in Table 3.2 (pp. 18–20).

## PROBLEM SET

**1.** Polytetrafluoroethylene has a molecular weight of $2 \times 10^6$ g/mol.
   **(a)** What is the repeating unit?
   **(b)** What is the molecular weight of the repeating unit?
   **(c)** What is the degree of polymerization?

**2.** Define the following terms:
   **(a)** Polymer               **(f)** Block copolymer
   **(b)** Linear polymer        **(g)** Degree of polymerization
   **(c)** Branched polymer      **(h)** Monomer
   **(d)** Copolymer             **(i)** Crosslinked polymer
   **(e)** Random copolymer

**3.** Using the symbols —A— and —B— to designate two types of repeating units, draw representative segments of (b) through (f) in Problem 2.

**4.** Polymethylmethacrylate (Plexiglas* or Lucite**) has the following structure:

Is this an example of a branched polymer? Why or why not?

*Plexiglas is a Rohm and Haas trademark.

**Lucite is a du Pont trademark.

5. A polyvinyl chloride (PVC) molecule has a degree of polymerization of 30,000.
   (a) What is the repeating unit for PVC?
   (b) What is the molecular weight of the repeating unit?
   (c) What is the molecular weight of the polymer?

6. F. Rodriguez's *Principles of Polymer Systems* [4], G. Odian's *Principles of Polymerization* [5], and K. J. Saunders's *Organic Polymer Chemistry* [1] are good references for some of the chemical work involving polymers. Use these sources to find and draw the monomers for the polymers listed in Table 3.2.

7. Each January, *Modern Plastics* magazine publishes a summary of the production volumes and major areas of application for synthetic polymers. Also included are prices for the various polymers. Obtain an up-to-date copy and compare the costs per pound of the various polymers listed in Table 3.2.

8. Using the *Modern Plastics Encyclopedia*, find a supplier for each of the polymers listed in Table 3.2.

9. If vinyl chloride is a carcinogenic material, why is pure PVC not carcinogenic also?

It is assumed in this text that the student already has some background in statics, dynamics, and strength of materials. Since the basic principles of these courses will be used repeatedly in this text, students are encouraged to do the following problems as a review.

10. The following is a typical stress–strain curve for a semicrystalline polymer such as low-density polyethylene:

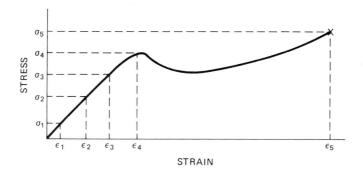

Using this curve, identify or write expressions for the following:
   (a) Elastic modulus
   (b) Yield strength
   (c) Ultimate strength
   (d) Elongation to break
   (e) Toughness
   (f) Yield strain
*Example*:

$$\text{Resilience} = \text{Area under the elastic portion of the stress–strain curve} = 1/2\sigma_3\epsilon_3$$

See ASTM D 638 [6] for definitions of any of the above terms.

**11.** Distinguish between strength and elastic modulus.

**12.** On a stress–strain curve, illustrate the general behavior of the following types of materials:

    **(a)** Rigid, brittle, and strong

    **(b)** Rigid, tough, and strong

    **(c)** Rigid, brittle, and weak

    **(d)** Flexible, tough, and strong

**13.** Terms such as "rigid" and "flexible" are relative. That is, their meanings depend on the type of material being discussed. For example, polycarbonate is considered a rigid, tough, and strong polymer. Use the *Modern Plastics Encyclopedia* to obtain the tensile strength, modulus, and elongation to break for polycarbonate in an unmodified form. How do these values compare to typical values for low-carbon steel?

**14.** The following is a load versus elongation curve for a semicrystalline polymer:

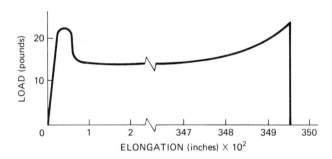

The specimen geometry was as follows:

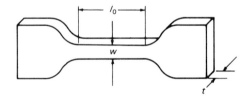

$l_0$ = GAGE LENGTH = 0.709 IN.

$t$ = THICKNESS = 0.030 IN.

$w$ = WIDTH = 0.1875 IN.

Calculate:

**(a)** Tensile modulus

**(b)** Proportional limit strength

**(c)** Yield strength

**(d)** Ultimate strength

**(e)** Elongation to break

Report these values in both English and SI units.

## REFERENCES

[1] Saunders, K. J., *Organic Polymer Chemistry*, Chapman and Hall, London (1973).

[2] Mark, H. F., Polymer Chemistry: The Past 100 Years, *Chemical and Engineering News* (April 6, 1976), p. 176.

[3] ———, Materials 1981, *Modern Plastics*, vol. 58, no. 1, p. 67.

[4] Rodriguez, F., *Principles of Polymer Systems*, McGraw-Hill, New York (1970).

[5] Odian, G., *Principles of Polymerization*, McGraw-Hill, New York (1970).

[6] American Society for Testing and Materials, *ASTM Standards, Part 35*, Philadelphia (1974).

**TABLE 3.2 Repeating Units of Common Commercial Polymers**

| Polymer | Repeating Unit | Typical Applications |
|---|---|---|
| Polyethylene (PE) | $\left[\begin{array}{c} \text{H} \quad \text{H} \\ -\text{C}-\text{C}- \\ \text{H} \quad \text{H} \end{array}\right]_n$ | Vapor barriers, tanks, packaging films, wire and cable insulation |
| Polypropylene (PP) | $\left[\begin{array}{c} \text{H} \quad \text{CH}_3 \\ -\text{C}-\text{C}- \\ \text{H} \quad \text{H} \end{array}\right]_n$ | Tanks, textile fiber, rope |
| Polyvinyl chloride (PVC) | $\left[\begin{array}{c} \text{H} \quad \text{Cl} \\ -\text{C}-\text{C}- \\ \text{H} \quad \text{H} \end{array}\right]_n$ | Pipe, valves, fittings, flooring, wire and cable insulation |
| Polytetrafluoroethylene (PTFE) | $\left[\begin{array}{c} \text{F} \quad \text{F} \\ -\text{C}-\text{C}- \\ \text{F} \quad \text{F} \end{array}\right]_n$ | Seals, valve components, coatings |
| Polystyrene (PS) | (repeating unit with pendant benzene ring) | Insulating foams, lighting panels, appliances, furniture components |

## TABLE 3.2 (Continued)

| Polymer | Repeating Unit | Typical Applications |
|---|---|---|
| Nylon 6 | | Bearings, gears, cams, tire cord, textile fiber, rope |
| Nylon 6/6 | | Automotive bodies and boat hulls, tire cord, textile fiber |
| Polyester | | Insulating and cushioning foams, coatings, adhesives |
| Polyurethane | | Seals, coatings, adhesives |
| Polydimethyl siloxane (silastic rubber) | | Tires |
| cis-1,4-polyisoprene (natural rubber) | | Bearings, gears |
| Polyformaldehyde | | |

19

**TABLE 3.2 (Continued)**

| Polymer | Repeating Unit | Typical Applications |
|---|---|---|
| Polyacrylonitrile (PAN) | | Textile fiber |
| Polyphenylene oxide | | Automotive and electrical components |
| Polysulfone | | Electrical components |
| ABS | A mixture of polybutadiene (see Chapter 14) particles dispersed in and bonded to a matrix of styrene acrylonitrile copolymer | Pipe, automotive components |
| Epoxy | See Chapter 16 | Adhesives, coatings |
| Polymethyl methacrylate | | Glazing, coatings |
| Phenol formaldehyde | | Adhesives |

**TABLE 3.2 (Continued)**

| Polymer | Repeating Unit | Typical Applications |
|---|---|---|
| Urea formaldehyde | | Adhesives |
| Melamine formaldehyde | | Adhesives |
| Polycarbonate (from bisphenol A) | | Electrical components |

*Note:* An "R" or "R'" in the repeating unit indicates that a number of organic structural units can be present at that position. For example, in the polyester repeating unit "R" could be $(CH_2)_2$ or $(CH_2)_4$, and "R'" could be a benzene ring. Both are polyesters because of the $-\overset{\displaystyle O}{\underset{\displaystyle \|}{C}}-O-$ linkage.

As mentioned in the first chapter, the bonds between polymer molecules are usually van der Waals forces, permanent dipoles, or hydrogen bonds. These are in part responsible for many of the properties observed. In some cases it is desirable to supplement this intermolecular bonding with covalent bonds (or covalently bonded systems) between otherwise individual chains. These bonds are called *crosslinks*; they are important in polymeric products such as styrene butadiene rubber (for tires), epoxy and phenolic adhesives, polyurethane foams, and some acrylic coatings. Crosslinking methods and consequences of crosslinking are considered in this chapter.

## CROSSLINKING METHODS

Crosslinks can be formed by several methods. One of these is with the use of monomer units having a functionality greater than 2. This was the case with the MMA/TMPTMA system mentioned earlier, and it is also the case with other polymers such as epoxies, phenolics, melamines, and ureas. The structures of these polymers, together with their monomers, are shown in Chapter 16. In these cases the crosslinks form as the main chain forms, and the resulting network is actually one molecule with branches and main chain indistinguishable from one another.

Crosslinks can also be formed in materials that are already polymeric, by using various chemical agents. This process is sometimes referred to as curing. For example, in natural rubber (*cis*–1,4–polyisoprene):

$$\left[\begin{array}{ccccc} & H & CH_3 & H & H \\ & | & | & | & | \\ -& C-C & =C & -C- \\ & | & & & | \\ & H & & & H \end{array}\right]_n$$

sulfur is added to the polymer during the compounding process. When the material is shaped and heated, the sulfur attacks the reactive sites on the rubber molecules. The reactions that occur are complex (see [1]), but one possibility is the following:

$$\sim\left[\begin{array}{c}H \;\; CH_3 \;\; H \;\; H\\ |\quad|\quad|\quad|\\ C-C=C-C\\ |\qquad\qquad|\\ H\qquad\qquad H\end{array}\right]_x \!\!\underset{(S)}{}\!\! \left[\begin{array}{c}H \;\; CH_3 \;\; H \;\; H\\ |\quad|\quad|\quad|\\ C-C=C-C\\ |\qquad\qquad|\\ H\qquad\qquad H\end{array}\right]\left[\begin{array}{c}H \;\; CH_3 \;\; H \;\; H\\ |\quad|\quad|\quad|\\ C-C=C-C\\ |\qquad\qquad|\\ H\qquad\qquad H\end{array}\right]_y \!\!\sim$$

$$\sim\left[\begin{array}{c}H \;\; CH_3 \;\; H \;\; H\\ |\quad|\quad|\quad|\\ C-C=C-C\\ |\qquad\qquad|\\ H\qquad\qquad H\end{array}\right]_x \!\!\underset{H}{(S)}\!\! \left[\begin{array}{c}CH_3 \;\; H \;\; H\\ |\quad|\quad|\\ C-C=C-C\\ |\qquad\qquad|\\ \qquad\qquad H\end{array}\right]\left[\begin{array}{c}H \;\; CH_3 \;\; H \;\; H\\ |\quad|\quad|\quad|\\ C-C=C-C\\ |\qquad\qquad|\\ H\qquad\qquad H\end{array}\right]_y \!\!\sim$$

This process, which is also known as vulcanization, was discovered by Charles Goodyear in 1839.

Sulfur is also used to crosslink other rubbery polymers such as styrene butadiene rubber (SBR), acrylonitrile–butadiene (nitrile) rubber, and polybutadiene rubber. Chemicals other than sulfur can also be used to crosslink specific systems. Some of these crosslinked materials are rubbery, and some are not. For example, peroxides are used to crosslink rigid polyethylene* as well as rubbery ethylene–propylene copolymers. The reader is encouraged to explore references [1] and [2], listed at the end of this chapter, for details.

A third method of forming crosslinks in polymers is with nuclear radiation. Although radiation crosslinking is used less extensively than the two methods just described, it is nonetheless an important consideration in several specific systems. For example, electron-beam radiation is utilized extensively to crosslink wire and cable insulation. Production of heat-shrinkable tubing and film for a variety of applications also involves crosslinking with electron beams.

In general, nuclear radiation produces free radicals and ions by various intermediate processes depending on the incoming nuclear particle and its energy. Nuclear particles utilized include photons (from the ultraviolet range through the energies of X-rays and gamma rays), neutrons, energetic electrons, protons, and alpha particles. The radiation-induced radicals and ions are the initiators of subsequent reactions including chain degradation, gas formation, and crosslinking, which can occur simultaneously at varying rates. When crosslinking and degradation occur simultaneously, for instance, the relative rate of each and thus the dominant process depends on such factors as the polymer repeating unit, the atmosphere, and the temperature.

---

*Most polyethylene is not crosslinked. Some grades are, however, in order to obtain specific property improvements such as an increase in the upper temperature use limit.

    In certain cases the effects radiation has on polymer properties are undesirable. This occurs, for example, in radiation sterilization of biomedical devices. Here the primary purpose of the radiation is to kill bacteria on the polymer that could cause infection. At the same time, however, the radiation (usually gamma) will change the properties of the polymer through crosslinking and/or degradation effects. In such cases these effects are undesirable, and polymers that are very susceptible to radiation degradation may require other means of sterilization.

## GENERAL EFFECTS OF CROSSLINKS
## ON PROPERTIES OF POLYMERS

In examining the influence of crosslinks (or any other characteristic of polymers) on bulk properties, it is always helpful to try to visualize what is occurring on a molecular level. In crosslinking, for example, assume that initially a disordered, uncrosslinked polymer exists:

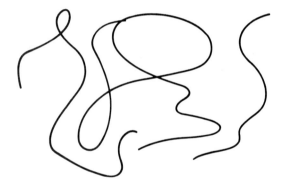

Crosslinking will create points in this structure where portions of individual molecules are now rigidly fixed in position relative to each other:

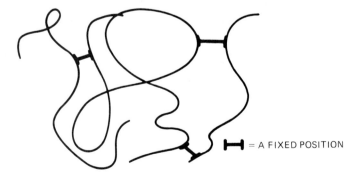

                                                             = A FIXED POSITION

This occurs because the secondary bonds at these positions have been replaced by much stronger covalent bonds. The net result is a more rigid network, which in turn is reflected in changes in the bulk properties of the material. In tensile behavior, for example, the addition of crosslinks generally results in an increase in modulus and a decrease in the strain to break.

Crosslinking necessarily affects all properties that are influenced by the type and degree of intermolecular bonding. Solubility, for example, is eliminated because the chains can no longer be separated by the solvent molecules. The polymer can swell, but it cannot dissolve. Similarly, the ability of the polymer to melt or soften and become a liquid is eliminated. Properties such as thermal expansion, conductivity, heat-distortion temperature, hardness, impact strength, and creep are also affected. These influences will be explained in detail in the chapters dealing with these specific properties.

## THERMOPLASTICS AND THERMOSETS

A major classification or distinction in polymers is based on the presence or absence of crosslinks. In earlier times (before much was known about the structures of polymers) two types of thermal behavior were observed. First, there were polymers that could be heated to a softening point, shaped by pressure, and cooled to retain that shape. This could be done repeatedly. Second, there were materials that could be heated to a softening point, shaped by pressure, and, if desired, removed from the hot mold without cooling. This process could not be repeated; additional heat and pressure only led to degradation. Materials with the first type of behavior were called thermoplastics, while those with the latter behavior were called thermosets. This terminology is still used today. In thermoplastics only secondary van der Waals forces, dipoles, and hydrogen bonds exist between the chains; there are no crosslinks or crosslinkable points that can crosslink upon the first application of heat. In a thermoset, however, heat-crosslinkable sites exist before its first exposure to heat, and these sites form crosslinks during the heating process. Once formed, the crosslinks hold the shape of the article. They allow it to be pulled hot from the mold and resist any further change in shape.

In a general sense a thermoplastic can be considered a noncrosslinkable polymer, while a thermoset is a crosslinkable or crosslinked one. The bit of confusion in this lies in the fact that the thermoset will flow with heat and pressure (while it is a low-molecular-weight system) before the crosslinks are formed. To summarize:

THERMOPLASTICS                                  THERMOSETS

LINEAR                                          CROSSLINKED

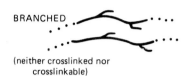

THERMOPLASTICS

BRANCHED

(neither crosslinked nor
crosslinkable)

THERMOSETS

CROSSLINKABLE (will crosslink
upon processing)

To confuse the issue more, there are a few uncrosslinked polymers that nonetheless behave as though they are cured thermosets; that is, when they are heated, they will degrade rather than soften and flow. A few polymers such as polyacrylonitrile and cellulose exhibit this behavior. In each case the secondary bonds (while individually weaker than the primary bonds that make up the chain) exist in such great numbers that the thermal energy necessary to disrupt them exceeds the dissociation energy of the main chain bonds. In other words, what the secondary bonds in these examples lack in strength they more than make up in number. These are the exceptions rather than the rule, however, in the general classification of thermoplastics and thermosets.

## PROBLEM SET

1. Define:
    (a) Crosslink
    (b) Thermoplastic
    (c) Thermoset

2. What are the three basic methods by which crosslinks are formed? Which two are the most common?

3. Which of the methods in Problem 2 are involved in the crosslinking of the following polymers?
    (a) Epoxies
    (b) Natural rubber
    (c) Phenol formaldehyde adhesives
    (d) Crosslinkable polyethylene
    (e) MMA/TMPTMA

4. Why does crosslinking affect each of the following?
    (a) Solubility
    (b) Melt viscosity
    (c) Recyclability of scrap
    (d) Modulus
    (e) Elongation to break

5. If scrap loss is a disadvantage in fabricating thermosets, what can be an advantage of using thermosets (with regard to the process only)?

## REFERENCES

[1] Saunders, K. J., *Organic Polymer Chemistry*, Chapman and Hall, London (1973).

[2] Odian, G., *Principles of Polymerization*, McGraw-Hill, New York (1970).

A crystal or a crystalline region is a portion of a material in which the molecules or atoms are arranged in an ordered pattern as opposed to a random arrangement. In metals and ceramics it is usually not difficult to picture this. In ceramics, for example, the diamond lattice is a relatively simple structure consisting of tetrahedrally bonded carbon atoms[1], as shown in Figure 5.1. In a metal such as $\alpha$-iron the situation is equally well visualized, with the iron atoms arranged in a body-centered cubic (BCC) lattice [1], as shown in Figure 5.2.

## MORPHOLOGY OF POLYMER CRYSTALS

Crystals can also exist in many types of polymers. In these cases, however, the elements of the crystals are not atoms that are easily arranged, as in the case of the diamond and $\alpha$-iron examples, but are instead large molecules. An ordered array of these is not easy to visualize, and the morphology of polymers crystallized under different conditions is still being studied.

The earliest concept of how large molecules formed crystals was the fringed micelle model. In this, portions of the polymer chains were visualized as aligning with one another in small regions (crystallites). These aligned regions were separated by regions of nonaligned or amorphous polymer molecules. Diagrammatically, this was represented as shown in Figure 5.3. By X-ray techniques the ordered regions were shown to be relatively small, which implied that each molecule was part of several crystallites and amorphous regions. This model, while simplistic, was useful for years in explaining some of the effects of drawing and other mechanically related phenomena that are observed in polymers.

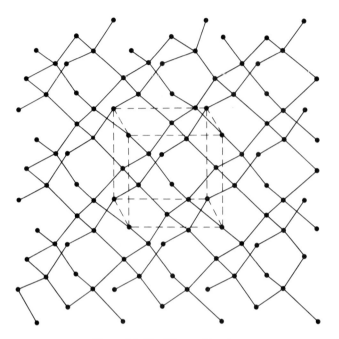

**Figure 5.1** The Diamond Lattice.

**Figure 5.2** The α-Iron Lattice.

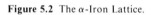

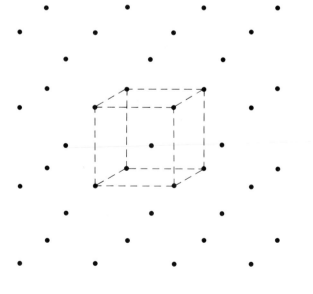

**Figure 5.3** Fringed Micelle Model of Polymer Crystallinity.

Later studies of polymer crystals showed that they are much more complex than the fringed micelle model suggests. In many polymers, lamellar structures (platelets) consisting of aligned, looped molecules have been identified. In these structures the molecules fold back on themselves and are aligned so that their lengths are perpendicular to the large faces of the platelets. This is illustrated in Figure 5.4.

Although these lamellar crystals were originally discovered as single crystals grown from dilute polymer solutions, they have also been found as components of larger ordered structures in melt-crystallized bulk polymers. The nature of these structures depends on the crystallization conditions. When polyethylene, for example, is crystallized from the melt under no external stress, structures such as those illustrated in Figure 5.5 result. These are called spherulites and, in aggregate, are somewhat analogous to the grain structure in metals. Spherulites form by crystallization that begins at an impurity or site of local order and then proceeds in all directions until other growing spherulites are encountered [2, 3]. Within each spherulite one of the crystal axes of the platelets is usually aligned radially or at some characteristic angle to the radii. In some forms of polyethylene, for example, the b axes (see Figure 5.5) are oriented along the radii of the spherulite and the a and c axes vary in a helical fashion about the radii [2].

**Figure 5.4** Lamellar Structure of a Polymer Crystal. (*Adapted from* [2].)

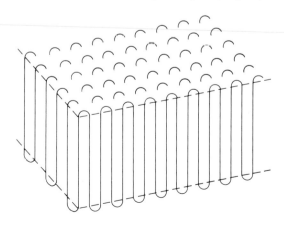

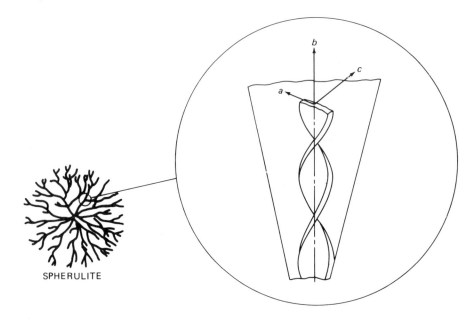

**Figure 5.5** Crystal Axis Orientation in a Polyethylene Spherulite (*C* axis is parallel to the backbones of the molecules). (*Adapted from* [2].)

In contrast to spherulites, a more fibrous type of structure can result when stresses exist during crystallization. Structures strongly resembling the shape of shish kebab have been observed under these conditions. A model of one of these is illustrated in Figure 5.6. The platelets in these structures are interconnected with bundles of parallel molecular segments. These segments are fibrous in nature and are aligned in the direction of the applied stress [4]. One would therefore expect relatively high tensile strength and modulus in the alignment direction when these structures exist.

It is apparent from Figures 5.5 and 5.6 that the crystalline morphologies of polymers are rather complex. Much remains to be learned in regard to these structures. An important point to note, however, is that polymers are not 100 percent crystalline. The spherulitic structure in Figure 5.5, for example, has significant amorphous components associated with the chain fold surfaces of the platelets, chain ends, spherulite boundaries, etc. Bulk polymers are never as highly crystalline as metals, in which the defects typically represent a very small percentage of the structure.

Polymers can also vary widely in the amount of crystallinity they possess. Those possessing no crystallinity are termed amorphous, while those that are not amorphous are termed semicrystalline (or crystalline for short, with the understanding that significant amorphous fractions also exist). Some common semicrystalline polymers are the polyethylenes, polytetrafluoroethylene, polypropylene, polyformaldehyde, the nylons, and the linear polyesters.

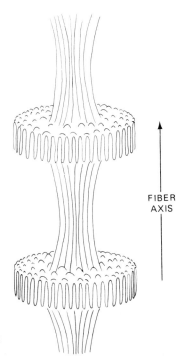

FIBER
AXIS

**Figure 5.6** Model of a "Shish Kebab"
Structure Resulting from Crystallization
under Stress.

One can sometimes estimate whether a polymer is crystalline or not simply by looking at it. Since the refractive indices of crystalline and amorphous regions are usually different, the light that passes through a crystalline polymer is scattered by the crystallites, and the object appears translucent. In an amorphous polymer there are no abrupt internal boundaries as far as refractive index is concerned, so the material tends to appear transparent (Figure 5.7). There are exceptions to these generalizations: These include situations in which (1) there is any type of filler in the material that obstructs light; (2) the crystallites are very

**Figure 5.7** Light Passing through Semicrystalline and Amorphous Polymers.

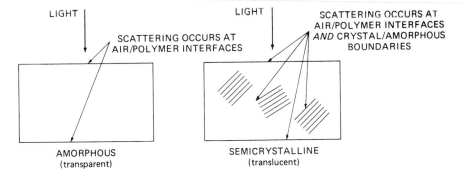

LIGHT

SCATTERING OCCURS AT
AIR/POLYMER INTERFACES

LIGHT

SCATTERING OCCURS AT
AIR/POLYMER INTERFACES
*AND* CRYSTAL/AMORPHOUS
BOUNDARIES

AMORPHOUS
(transparent)

SEMICRYSTALLINE
(translucent)

small in size or in number and do not interfere appreciably; and (3) the material is relatively thin and light is therefore not scattered to a great extent (such as in polyethylene film, or Saran* wrap).

## FACTORS AFFECTING CRYSTALLINITY

That some polymers are crystalline and others are not is no accident. Two fundamental requirements must be met in order for crystallization to be a possibility. These involve intermolecular bond strength and the regularity of the molecular structure [5] and are discussed in the sections that follow.

### Intermolecular Bond Strength

For crystals to exist at a given temperature, the intermolecular forces must be sufficient to overcome the disordering tendency of thermal vibration. Conversely, in order for a crystal to melt, the thermal vibrational energy must overcome the intermolecular bonds. Strongly bonded crystals require relatively high temperatures to disrupt them, and weakly bonded crystals may be liquids at room temperature. The same argument also applies to gases, liquids, and solids of low molecular weight. To crystallize a liquid or a gas, it must be cooled to the point where the intermolecular bonds predominate over the disordering tendency of the thermal vibrations.

### Regular Chain Structure

To have an ordered arrangement of molecules, the molecules themselves must have a rather regular, ordered structure. For example, the presence of random short-chain branching in some polymers correlates with lower crystallinity. In the polyethylenes there is a tendency for short branches (typically ethyl or butyl groups) to form along the chain. In low-density polyethylene (LDPE) these are fairly numerous (15 to 30 branches per 1000 carbons), whereas in high-density polyethylene (HDPE) they are less frequent (1 to 5 branches per 1000 carbons). The result is that in the case of LDPE one is trying to align chains that look as follows:

and in the case of HDPE, as follows:

*Saran is a Dow Chemical Co. trademark.

One therefore finds that HDPE molecules align to a higher degree than LDPE molecules. The most obvious result of higher crystallinity in polyethylene is higher density. HDPE has a density between 0.941 and 0.965 g/cm³, whereas LDPE has a density from 0.910 to 0.925 g/cm³ [6] because of the closer packing (higher crystallinity) of the less branched material.

The presence of chlorine, fluorine, methyl groups, etc., on the backbone carbon atoms can also influence the geometrical regularity of the molecular structure. Since these atoms and molecular groups are larger than hydrogen (see Figure 5.8), under some circumstances they can interfere with the alignment process.

More is involved than just size, however. If one considers a three-dimensional representation of an uncoiled polymer chain, this becomes more evident. Because of the 109.5° tetrahedral bonding nature of the carbon atom, a zigzag chain results:

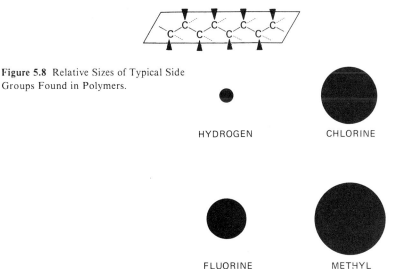

**Figure 5.8** Relative Sizes of Typical Side Groups Found in Polymers.

HYDROGEN

CHLORINE

FLUORINE

METHYL

BENZENE RING

The two carbon bonds not forming the backbone of the molecule project above (heavy line) and below (dashed line) the plane of the other bonds as shown. These are the bonds to which the hydrogens, chlorines, fluorines, methyl groups, etc. are attached in different types of polymers. In polypropylene,

$$\left[\begin{array}{c} H \quad CH_3 \\ -C-C- \\ H \quad H \end{array}\right]_n$$

this three-dimensional planar zigzag form can occur in three completely different ways. Depending on the relative positions of the methyl groups, either isotactic, syndiotactic, or atactic polypropylene results. These three polymers are illustrated in Figure 5.9.

These three forms of polypropylene are fundamentally different from one another. One cannot change atactic polypropylene into isotactic polypropylene (or any other form into an alternative form) without breaking and rearranging some chemical bonds. Rotation and twisting about these bonds will not

**Figure 5.9** Isotactic, Syndiotactic, and Atactic Forms of Polypropylene.

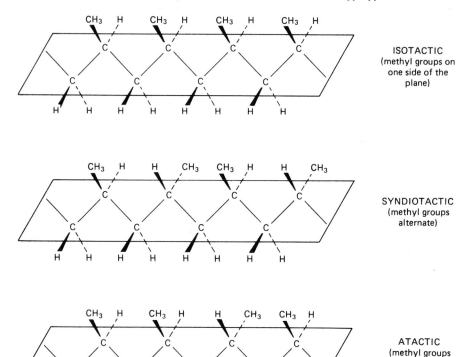

ISOTACTIC
(methyl groups on
one side of the
plane)

SYNDIOTACTIC
(methyl groups
alternate)

ATACTIC
(methyl groups
randomly attached
to each side)

accomplish this. These structures illustrate stereoisomerism, which involves the asymmetric carbon atom:

$$W-\underset{\underset{Z}{|}}{\overset{\overset{X}{|}}{C}}-Y$$

An asymmetric carbon atom is one to which four different groups are bonded. In a polymer two of the substituent groups on an asymmetric carbon atom located in the backbone of the molecule are the remaining segments of the polymer chain.

Similar arguments can be made for all other polymers containing asymmetric carbon atoms. Many of these (like polypropylene) are based on the vinyl structure:

$$\left[\begin{array}{cc} H & H \\ | & | \\ C & -C \\ | & | \\ H & X \end{array}\right]$$

where X can be any atom or group except hydrogen. These include polystyrene, polyvinyl chloride, polyvinyl acetate, polyvinyl alcohol, and polyacrylonitrile.

The significance of this is that a regular structure has a greater tendency to crystallize than an irregular one. Thus the isotactic and syndiotactic structures tend to be more crystalline than the atactic forms. The molecules that are either isotactic or syndiotactic are alike (except for chain length) and can better fit together than those in the atactic case. In the atactic form of polypropylene, for example, the methyl groups act like short chain branches, appearing on one side or the other in a random fashion. This makes it difficult to find two regions of chains that are similar enough to align. As a result, the crystallinity is low in atactic polypropylene and the material is sticky, soft, and weak. The stereoregular (isotactic) form, however, is highly crystalline, strong, and hard. Atactic polypropylene has few commercial applications, but isotactic polypropylene is one of higher-volume synthetic polymers in use today.

## THE INFLUENCE OF FABRICATION CONDITIONS ON CRYSTALLINITY

Polymer crystals tend to be sensitive to the conditions under which they are formed, whether in a laboratory or on a production line. Chain mobility in the melt or in a solution is much less than that of smaller molecules or atoms; thus a significant amount of time can be required in certain situations for these molecules to align and crystallize. Many polymers, for example, can be quenched from the melt to form completely amorphous systems. This is routinely done in the production of soft drink bottles using thermoplastic polyesters. The low thermal conductivity of polymers (approximately one thousandth that of metals) makes them very susceptible to the effects of the cooling rate. In a thermoplastic molding process involving a reasonably thick part, for example, the center of the part will cool at a much slower rate than the outer portion, which is in direct contact with the mold. Thus one would expect a different degree and type of crystallinity on the surface from that in the core. The degree to which this occurs depends on the individual polymer.

Crystallinity is also influenced by stress, both during crystallization (as mentioned previously) and afterwards. This is the basis, for example, of the drawing process by which the mechanical properties of fibers are enhanced. By stretching the fibers one can induce more of the molecular segments to align, and also to align in the direction of the applied stress. This results in the relatively high strength and modulus in the fiber direction that are required for many applications.

Sometimes, however, alignment due to stress in fabrication occurs when it is not desired. While it is not of the same magnitude as that which occurs in the drawing of fibers, it can result in an undesirably anisotropic product. In the extrusion of thick film or sheet, for example, the stresses placed on the long molecules as they are forced through the die tend to align them in the extrusion direction. The result is an oriented product, which has different properties (such as tear resistance) in different directions.* Various methods such as crosslaminating and biaxial orientation have been used to circumvent this problem.

## THE INFLUENCE OF CRYSTALLINITY ON PROPERTIES

The importance of crystallinity in determining bulk properties of polymers cannot be overemphasized. Many properties such as modulus, strength, elongation to break, impact strength, conductivity, solubility, and hardness are influenced by crystallinity. In a certain sense their effects are similar to those of crosslinks, since they provide relatively rigid regions in the molecular structure. They are not the same, however, and the relationship between crystallinity and bulk properties is a rather complex one.

The polyethylenes provide a nice illustration of the effects of crystallinity changes on various properties for a given polymer. As noted earlier, the degrees of short-chain branching in high- and low-density polyethylene are different; this in turn influences the amount of crystallinity. Table 5.1 compares some pertinent properties of these two types of polyethylene. The reasons for these property differences will be examined in greater detail in later chapters. It should be emphasized, however, that despite the higher crystallinity and concomitant higher strength and modulus of high-density polyethylene, there are applications

---

*Orientation can also occur in amorphous materials. See Nielsen [7] for details concerning the effects of this on mechanical properties.

TABLE 5.1 Typical Properties of Polyethylenes

| Property | Low-Density | High-Density |
|---|---|---|
| Density $(g/cm^3)$ | 0.910–0.925 | 0.941–0.965 |
| Crystal melt temperature (°C) | 95–130 | 120–140 |
| Tensile strength (MPa)* | 4.1–15.9 | 21.4–37.9 |
| Tensile modulus (MPa)* | 96.5–262 | 414–1250 |
| Elongation to break (%)* | 90.0–800.0 | 20–1300 |
| Hardness (Shore) | D41–D50 | D60–D70 |

*ASTM D 638

(*Adapted from* [6].)

for both types of polyethylene. LDPE, for example, is used extensively in packaging films because of its relative clarity, flexibility, and toughness. HDPE, on the other hand, has sufficient crystallinity to impart the hardness, stiffness, and strength needed in such items as rotationally molded gasoline tanks. In other words, one does not always want the hardest and strongest material for a particular job. There are usually a number of considerations other than these two properties. An indication of this is that a significantly larger volume of LDPE than of HDPE is produced in this country.

## PROBLEM SET

1. What two basic requirements must be met in order for crystallization to be possible in polymers? Which of these is missing in the following examples?
   (a) Polyethylene at 200°C
   (b) Atactic polyvinyl chloride at room temperature
   (c) Isotactic polystyrene dissolved in acetone
   (d) A random ethylene–propylene copolymer at room temperature
2. What influence can drawing have on the amount and orientation of crystallinity? What major product relies on this effect?
3. On a planar zigzag drawing such as the following

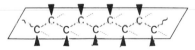

   draw the isotactic, syndiotactic, and atactic configurations of polypropylene, polystyrene, and polyvinyl chloride.
4. Polystyrene is typically transparent. What does this indicate about its probable tacticity?
5. Is it possible to change isotactic polypropylene into atactic polypropylene by twisting or rotating the bonds?
6. Explain why a semicrystalline polymer is usually translucent while an amorphous one might not be. What could prevent an amorphous polymer from being transparent?

7. How is density influenced by crystallinity? Why?

8. What are some applications in which crystallinity cannot be tolerated?

9. In what basic way are crosslinks and crystals similar? What are some ways in which they are different?

## REFERENCES

[1] Evans, R. C., *An Introduction to Crystal Chemistry*, Cambridge University Press, Cambridge (1964).

[2] Stein, R. S., and A. V. Tobolsky, Crystallinity in Polymers, in *Polymer Science and Materials* (A. V. Tobolsky and H. F. Mark, eds.), Wiley-Interscience, New York (1971).

[3] Keller, A., Organization of Macromolecules in the Solid State; A Personal Approach, *Journal of Polymer Science*, Symposium No. 51 (1975), pp. 7–44.

[4] Keller, A., Polymer Crystallization: A Survey and Salient Current Issues, *Journal of Polymer Science*, Symposium No. 59 (1977), pp. 1–10.

[5] Rosen, S. L., *Fundamental Principles of Polymeric Materials*, Wiley, New York (1981).

[6] *Modern Plastics Encyclopedia*, McGraw-Hill, New York (1980–1981).

[7] Nielsen, L. E., *Mechanical Properties of Polymers and Composites*, vols. 1 and 2, Dekker, New York (1974).

# molecular weight

## chapter 6

## MOLECULAR WEIGHT AVERAGES

It was noted in an earlier chapter that the length of the polymer chains has a strong influence on the properties of the material. Because of the random method by which the polymer chains grow and terminate their growth (Chapter 16), a distribution of molecular lengths almost always occurs in bulk polymers. This can be represented by the following graph:

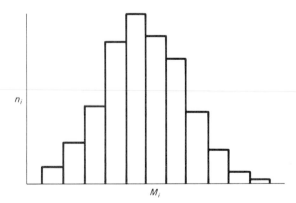

where $n_i$ is the number of molecules with molecular weight $M_i$. If one is to associate a particular molecular weight or change in molecular weight with a property or change in a property, then a method of characterizing the distribution of molecular weights is required. This involves the use of molecular weight averages.

The term "average" generally suggests a numerical average in which the sum is determined and divided by the total number of elements contributing to that sum. For example, if the following group of people were on a football team:

| Number of People | Weight (kg) |
|:---:|:---:|
| 1 | 50 |
| 2 | 70 |
| 2 | 90 |
| 1 | 140 |

their average weight would be

$$\frac{50 + 70 + 70 + 90 + 90 + 140}{6} = 85 \text{ kg}$$

and the heaviest players would be considered above average and the lighter players would be considered below average in weight.

One can also solve this problem using a slightly different approach. Since there are 6 people, those weighing 50 kg represent one sixth of the total number; those weighing 70 kg represent two sixths of the total number; those weighing 90 kg represent two sixths of the total number; and those weighing 140 kg represent one sixth of the total number. The average is

$$\tfrac{1}{6}(50) + \tfrac{2}{6}(70) + \tfrac{2}{6}(90) + \tfrac{1}{6}(140) = 85 \text{ kg}$$

The answer is the same, since the two methods are equivalent. In either case this average is called the *number average* because it is based on counting or weighting each person equally.

It is not necessary to count these people as equal in value for all purposes, however. In the case of a football team it may make more sense to count the heavier people more and the lighter people less in determining the average, since their effectiveness may depend on their weight. They might be counted instead in proportional to their contribution to the total weight of the group. For example, the lightest person would count as $\frac{50}{510}$ instead of $\frac{1}{6}$, and the heaviest person would count as $\frac{140}{510}$ instead of $\frac{1}{6}$. If an average is calculated using these values, the result is

$$\tfrac{50}{510}(50) + \tfrac{140}{510}(70) + \tfrac{180}{510}(90) + \tfrac{140}{510}(140) = 94 \text{ kg}$$

This average is called the *weight average*, since it counts each person in proportion to the fraction of the total weight he or she represents. The weight average is as valid as the number average; it is just a different approach with different applications.

It can be shown mathematically that the weight average is always greater than or equal to the number average. If the distribution of weights is monodisperse (i.e., everyone has the same weight) then the weight and number averages are equal and their ratio is 1. In all cases where a distribution of weights occurs (polydisperse), the ratio (weight average/number average) is greater than 1. As the distribution broadens, the ratio becomes larger.

Molecular weight averages for polymers are treated in a similar manner by substituting polymer molecules for people and molecular weights for the weights of people. For example, assume the following distribution of polymer molecules:

| $n_i$ = Number of Molecules | $M_i$ = Molecular Weight of ith Fraction |
|---|---|
| $6.02 \times 10^{23}$ | 10,000 g/mol |
| $12.04 \times 10^{23}$ | 15,000 g/mol |
| $12.04 \times 10^{23}$ | 20,000 g/mol |
| $6.02 \times 10^{23}$ | 30,000 g/mol |

The number average molecular weight $(\overline{M}_n)$ is

$$\overline{M}_n = \frac{\sum\limits_{i=1}^{n} n_i M_i}{\sum\limits_{i=1}^{n} n_i}$$

$$= \frac{(6.02 \times 10^{23})(10,000) + (12.04 \times 10^{23})(15,000) + (12.04 \times 10^{23})(20,000) + (6.02 \times 10^{23})(30,000)}{(6.02 \times 10^{23}) + (12.04 \times 10^{23}) + (12.04 \times 10^{23}) + (6.02 \times 10^{23})}$$

$$= 18,300 \text{ g/mol}$$

and the weight average molecular weight $(\overline{M}_w)$ is

$$\overline{M}_w = \frac{\sum\limits_{i=1}^{n} n_i M_i^2}{\sum\limits_{i=1}^{n} n_i M_i}$$

$$= \frac{(6.02 \times 10^{23})(10,000)^2 + (12.04 \times 10^{23})(15,000)^2 + (12.04 \times 10^{23})(20,000)^2 + (6.02 \times 10^{23})(30,000)}{(6.02 \times 10^{23})(10,000) + (12.04) \times 10^{23})(15,000) + (12.04 \times 10^{23})(20,000) + (6.02 \times 10^{23})(30,000)}$$

$$= 20,500 \text{ g/mol}$$

The relationship between these results and the football analogy can be seen after a few algebraic manipulations. In the number average,

$$\overline{M}_n = \frac{\sum\limits_{i=1}^{n} n_i M_i}{\sum\limits_{i=1}^{n} n_i}$$

the denominator is a constant for a particular distribution (the total number of molecules) and thus can be inserted inside the summation sign in the numerator:

$$\overline{M}_n = \sum_{i=1}^{n} \left( \frac{n_i M_i}{\sum\limits_{i=1}^{n} n_i} \right)$$

This is equivalent to the expression

$$\overline{M}_n = \sum_{i=1}^{n} \left( \frac{n_i}{\sum\limits_{i=1}^{n} n_i} \right) M_i$$

where $n_i / \sum_{i=1}^{n} n_i$ is the number or mole fraction (the $\frac{1}{6}$, $\frac{2}{6}$, etc. in the football example).

In a similar manner, $\overline{M}_w$ can be reduced as follows:

$$\overline{M}_w = \frac{\sum\limits_{i=1}^{n} n_i M_i^2}{\sum\limits_{i=1}^{n} n_i M_i} = \sum_{i=1}^{n} \left( \frac{n_i M_i^2}{\sum\limits_{i=1}^{n} n_i M_i} \right) = \sum_{i=1}^{n} \left( \frac{n_i M_i}{\sum\limits_{i=1}^{n} n_i M_i} \right) M_i$$

where the expression $(n_i M_i / \sum_{i=1}^{n} n_i M_i)$ is the weight fraction (the $\frac{50}{510}$, $\frac{140}{510}$, $\frac{180}{510}$, etc. in the football example).

The ratio $\overline{M}_w / \overline{M}_n$ is called the polydispersity index, or P.D.I. Typically, this ranges from 2 to 30 for commercial synthetic polymers. Specialized techniques can produce polydispersity indices close to 1, however.

## MEASUREMENT OF MOLECULAR WEIGHT AVERAGES

Every batch of polymer produced has a specific $\overline{M}_n$ and $\overline{M}_w$. These values reflect the size of the molecules and the distribution of chain lengths. To obtain these values, the manufacturer does not count each molecule individually and create a distribution like those described previously.* Instead, these values are usually

---

*A technique known as GPC (gel permeation chromatography) comes very close to this, however. Details concerning this method can be obtained from Rosen [1] and Rodriguez [2].

obtained indirectly by measuring specific properties that are directly related to $\overline{M_n}$ or $\overline{M_w}$. There are many techniques in use, and many of them involve dissolving polymers in solvents.

The exact technique used for a particular molecular weight average determination is a function of the type of average required and the approximate value. For example, $\overline{M_n}$ is determined from the measurement of properties that depend only on the number of molecules present, and not their sizes. The most common method for determining $\overline{M_n}$ is the measurement of the osmotic pressure of a polymer solution. This method is useful for molecular weights above approximately $10^4$ g/mol. Below this some of the polymer molecules are small enough to penetrate the membrane used in the apparatus to separate the polymer solution from pure solvent. This penetration results in errors. Therefore, for molecular weights below $10^4$ g/mol, other techniques (end group analysis, vapor pressure osmometry) are preferred. Table 6.1 summarizes several of the common methods used in the determination of $\overline{M_n}$ and $\overline{M_w}$. Descriptions of these techniques can be found in a number of detailed polymer science texts [1]–[4]. Other methods not listed in Table 6.1 also exist for measuring molecular weight averages, but their use is more limited. These include electron and X-ray microscopy, isothermal distillation, measurement of elevation of boiling point, measurement of depression of freezing point, osmodialysis, and ultracentrifugation.

The measurement of molecular weight by the methods listed in Table 6.1 can be time-consuming and costly. In addition, most of these techniques involve relatively sophisticated equipment. One method of circumventing some of these problems is through the use of $\overline{M_v}$, or the viscosity average molecular weight. Unlike the absolute methods listed in Table 6.1, the viscosity average molecular weight is not directly based on theory. Instead, it is related to previous experimental results. This may be unacceptable for some purposes, but the convenience and low cost of this method make it valuable for many practical applications. Numerically, $\overline{M_v}$ is greater than $\overline{M_n}$ and less than $\overline{M_w}$. The details of this widely used empirical method can be found in Rosen [1], Seymour and Carraher [3], and Flory [4].

**TABLE 6.1  Common Methods Used for the Determination of $\overline{M_n}$ and $\overline{M_w}$**

| Method | Average | Approximate Useful Molecular Weight Range |
|---|---|---|
| End-group analysis | $\overline{M_n}$ | Up to 20,000 g/mol |
| High-speed membrane osmometry | $\overline{M_n}$ | $2 \times 10^4$ to $2 \times 10^6$ g/mol |
| Vapor pressure osmometry | $\overline{M_n}$ | Up to $4 \times 10^4$ g/mol |
| Light scattering | $\overline{M_w}$ | $10^4$ to $10^7$ g/mol |

(*Adapted from* [1] *and* [3].)

## THE INFLUENCE OF MOLECULAR LENGTH ON PROPERTIES

The length of the molecules in a given polymer has a strong influence on the bulk properties of that polymer. One method of illustrating this influence is to consider how the properties of materials based on the paraffin structure,

$$H \left[ \begin{array}{cc} H & H \\ | & | \\ C & C \\ | & | \\ H & H \end{array} \right]_n H$$

change as $n$ increases. Table 6.2 summarizes the properties of materials with the above repeating unit that differ from one another only in the value of $n$. The nature of the secondary bonds between the molecules (in this example, van der Waals forces) in each case is the same.

In a general sense the influence of chain length for a given repeating unit can be represented graphically as

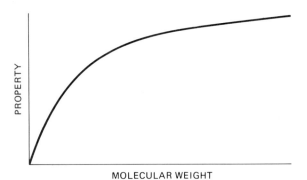

where initially there is a large change in the property in question (such as going from a gas to a liquid to a solid in the paraffin-structure example) followed by a diminished effect. This is true for properties such as modulus, tensile strength, impact strength, and resistance to creep.

One may be tempted to ask why all polymers are not made with very high molecular weights in order to take full advantage of increases in modulus, tensile strength, etc. In fact, most polymers do not have extremely high average molecular weights. With the exception of crosslinked polymers and a few thermoplastics (such as polytetrafluorethylene and ultrahigh-molecular-weight polyethylene), most polymers have number average molecular weights below 100,000 g/mol. Table 6.3 lists some typical values. A major reason for these relatively low values is that while one may improve the property of interest, other properties (particularly melt viscosity) may become unacceptable. If the molecular weight of a polymer is increased to obtain some property change, then melt viscosity will increase as well. It may then be necessary to process the

TABLE 6.2  The Influence of Molecular Weight on Materials Having
Paraffinlike Structures

| n | Molecular Weight | Material | Properties |
|---|---|---|---|
| 1 | 30 g/mol | Ethane | Gas at STP |
| 3 | 86 g/mol | Hexane | Liquid with a boiling point of 69° C |
| 20–34 | 562–954 g/mol | Paraffin | Waxy semisolid with a melting point of 50–55° C |
| ~700 | ~20,000 g/mol | Polyethylene | Fairly hard, tough solid with a melting point of 95–135° C |
| ~150,000 | ~4 × 10⁶ g/mol | Ultrahigh-molecular-weight polyethylene (UHMWPE) | Hard, tough solid with a melting point of 120–135° C and exceptional abrasion resistance |

polymer at a higher temperature in order to obtain the desired flow characteristics. This not only costs more but can also lead to degradation (see Chapter 9).

Most polymer systems are tradeoffs between the desired processability and
the end-use characteristics. In some cases it is possible to obtain the best of both
worlds. Polyethylene, for example, can be purchased in a crosslinkable form.
This is polyethylene with an added crosslinking agent. When it is processed by
melting (such as in an extruder or rotational molder), it flows with the
characteristics of its molecular weight. Meanwhile, the crosslinking process
begins and pushes the molecular weight up drastically. The result is a material
that processes as a low-molecular-weight polymer but has the finished properties
of one with higher molecular weight.

The influence which changes in molecular weight or molecular weight
distribution have on a particular property depend both on the property and on
the nature of the polymer. For example, molecular weight changes have a greater

TABLE 6.3  Typical Number Average Molecular Weights
of Some Commercial Polymers

| Polymer | $\overline{M}_n$ |
|---|---|
| LDPE | 20,000 |
| HDPE (Standard Oil process) | 15,000 |
| Nylon | 20,000 |
| Polyvinyl chloride | 40,000 |
| Polypropylene (Zeigler process) | 40,000 |
| Polyethylene terephthalate | 20,000 |

(*Adapted from* [5].)

influence on the melt viscosity of an amorphous polymer than they do on the modulus of that same polymer below its softening point. Similarly, some properties of polymers with hydrogen and permanent dipolar secondary bonds tend to reach the point of diminishing returns at lower molecular weights than do those of nonpolar polymers. The effects of molecular weight on specific properties will be considered in greater detail in later chapters.

## PROBLEM SET

1. Given the following molecular weight distribution, calculate $\overline{M}_n$, $\overline{M}_w$, and the polydispersity index:

| Molecular Weight | Number of Molecules |
|---|---|
| 5,000 | 2 |
| 15,000 | 4 |
| 25,000 | 6 |
| 50,000 | 2 |

2. Show that the equations for $\overline{M}_n$ and $\overline{M}_w$ as given in the text can also be expressed as

$$\overline{M}_n = \sum_{i=1}^{n} n_i M_i$$

where $n_i$ is mole fraction

$$\overline{M}_w = \sum_{i=1}^{n} w_i M_i$$

where $w_i$ is weight fraction

Repeat Problem 1 using these relationships.

3. Define the following terms:
   (a) Monodisperse polymer
   (b) Polydisperse polymer
   (c) Polydispersity index

4. Describe briefly the effect of increasing molecular weight on modulus and strength.

5. Given the following molecular weight distribution for PMMA,

| 10,000 g/mol | $5.25 \times 10^{24}$ molecules |
|---|---|
| 20,000 g/mol | 2.4 kg |
| d.p. = 300 | 0.5 mole |

   calculate $\overline{M}_n$ and $\overline{M}_w$. Recall that Total Weight ÷ Molecular Weight = Moles.

6. Why does chain length affect modulus?

**Answer:**

Modulus can be visualized to a certain extent as the resistance which segments of molecules have to sliding past one another.

This resistance depends on the length of the molecules as well as the type of intermolecular bonding. For chains (or chain segments) to slip, the forces between them must be disrupted. Long molecules require a larger force than small molecules because they have a larger number of intermolecular bonds per molecule.

In addition, the entwining of chains becomes important at higher molecular weights. As chain length increases, the degree of entanglement of the chains also increases. When these chains are stressed, the entanglements are temporary loops or knots in the system. They therefore retard slippage and increase the observed modulus.

7. Explain the rationale behind crosslinkable grades of polyethylene.
8. Why does chain length affect melt viscosity?

**Answer:**

Shear modulus in a solid and viscosity in a fluid are controlled to a large extent by the same factors. Therefore the explanation provided for Problem 6 also applies here. However, the degree of dependence on molecular weight is not the same for both properties. Above a critical value, melt viscosity is proportional to $(\overline{M_w})^{3.4}$ for many polymers [1]. This is in direct contrast to the diminishing-return effect that occurs with modulus at high molecular weights. This difference is thought to be due to the greater sensitivity of melt viscosity to chain entanglements.

9. What is the average molecular weight of a heavily crosslinked polymer?
10. Show mathematically that $\overline{M_w} \geq \overline{M_n}$.
11. Compare $\overline{M_n}$ and $\overline{M_w}$ for a mixture of an equal number of polymer molecules with $M_1 = 50,000$ g/mol, $M_2 = 100,000$ g/mol, and $M_3 = 200,000$ g/mol.
12. Compare $\overline{M_n}$ and $\overline{M_w}$ for a mixture of equal masses of polymer with $M_1 = 50,000$ g/mol, $M_2 = 100,000$ g/mol, and $M_3 = 200,000$ g/mol.
13. UHMWPE has a molecular weight far above that of most commercial polyethylenes.
    (a) What are some advantages of having this high molecular weight?
    (b) What are the disadvantages?
    (Suggested reference: [6], p. 74.)
14. What are two major differences between the molecular weights of polymers and those of compounds such as sucrose, hexane, and carbon tetrachloride?
15. Explain why one molecular weight average is not sufficient in describing the chain length of a polymer.
16. Using one or more of the appropriate references listed in the text, read and summarize the method for determining $\overline{M_n}$ with osmotic pressure techniques.
17. Repeat Problem 17 for $\overline{M_w}$ with light-scattering techniques.
18. Repeat Problem 17 for $\overline{M_v}$ with solution viscometry techniques.
19. What is melt index? How does it relate to molecular weight?

**Answer:**

Melt index refers to ASTM D 1238, "Measuring Flow Rates of Thermoplastics by Extrusion Plastometer" [7]. It is a measure of the rate at which a molten polymer is extruded under prescribed conditions through a die of a specified length and diameter. Numerically, melt index is the extrusion rate expressed in grams of polymer per 10 minutes. Melt index and molecular weight are inversely related; a polymer of high molecular weight will have a lower melt index than a polymer of lower molecular weight with the same structure. It is, therefore, a relatively quick and inexpensive method for comparing molecular weights. Commercial polymers (such as polyethylene and polypropylene) are usually available in a wide range of melt flow indices, since a range of values is necessary, in part to meet the requirements of different fabrication methods. Blow molding of polyethylene items, for example, usually requires a lower melt flow index than rotational molding because of the differing natures of the processes (see Chapter 18).

# REFERENCES

[1] Rosen, S. L., *Fundamental Principles of Polymeric Materials*, Wiley, New York (1981).

[2] Rodriguez, F., *Principles of Polymer Systems*, McGraw-Hill, New York (1970).

[3] Seymour, R. B., and C. E. Carraher, Jr., *Polymer Chemistry*, Dekker, New York (1981).

[4] Flory, P. J., *Principles of Polymer Chemistry*, Cornell University Press, Ithaca, N.Y. (1953).

[5] Saunders, K. J., *Organic Polymer Chemistry*, Chapman and Hall, London (1973).

[6] *Modern Plastics Encyclopedia*, McGraw-Hill, New York (1979–1980).

[7] American Society for Testing and Materials, *ASTM Standards, Part 35*, Philadelphia (1981).

properties of polymers

specific heat,
thermal conductivity,
and thermal expansion

chapter 7

## SPECIFIC HEAT

Specific heat is the energy required to raise the temperature of a unit mass one degree. The symbol for this property is $c$, and its units are $cal/g \cdot °C$ in the cgs system, $BTU/lb \cdot °F$ in the English system, and $J/kg \cdot K$ in the SI system. Table 7.1 lists the specific heats of several common materials.

It may seem confusing at first that a material such as copper or iron has a low specific heat compared to a material of much lower density, such as polyethylene. The figures are correct, however, since the units of specific heat are energy per gram or per pound, not per unit volume. It would obviously take more heat to warm a cubic meter of iron (approximately 7800 kg) than a cubic meter of polyethylene (approximately 1000 kg). The exact calculation is left as an exercise for the reader.

In considering the energy requirements for heating and melting polymers, it is important to remember that specific heat is a function of temperature. In transitionless regions it rises slightly with increasing temperature. At the crystal melt transition, however, a large change may occur (Figure 7.1) because of the latent heat of fusion. Polymers that lack a crystalline structure exhibit a discontinuity rather than a peak at their softening temperature ($T_g$).

## THERMAL CONDUCTIVITY

Given a material of cross-sectional area $A$ and thickness $X$, with a temperature gradient $\Delta T/X$, the heat flux under steady-state conditions is given by

**TABLE 7.1 Specific Heats of Some Common Materials (20° C)**

| Material | Specific Heat (J/kg · K) |
|---|---|
| Copper | 385 |
| Iron | 444 |
| Aluminum | 900 |
| Polytetrafluoroethylene | 938 (0° C rather than 20° C) |
| Polystyrene | 1194 |
| Polyformaldehyde | 1214 |
| Polymethyl methacrylate | 1388 |
| Nylon 6 | 1599 |
| Polypropylene | 1789 |
| High-density polyethylene | 1855 |
| Natural rubber | 1817 |
| Low-density polyethylene | 2315 |
| Ethanol | 2457 |
| Water | 4181 |

(*Adapted from* [1] *and* [2].)

**Figure 7.1** Temperature Dependence of the Specific Heat of a Semicrystalline Polymer.

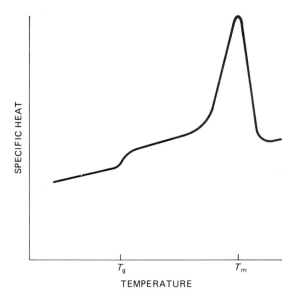

$$\text{Heat flux} = \frac{kA\ \Delta T}{X}$$

where $k$ is defined as the thermal conductivity.

Thermal conductivity is expressed in cal/s · cm · °C in the cgs system, BTU in/ft² · hr · °F in the English system, and W/m · K in the SI system. Table 7.2 is a summary of the thermal conductivities of some common materials.

As indicated in Table 7.2, polymers are generally poor conductors of heat. It is not difficult to visualize why this low conductivity occurs. For heat to flow, energy must be transferred by one or more of three mechanisms:

Convection:  by motion of mass from one point to another

Radiation:   via electromagnetic waves

Conduction: through a material or medium without translation of mass

In solids conduction is the primary mechanism, and this involves the transfer of thermal energy from one atom to another or from one molecule to another. This may be easy or difficult, depending on the nature of the bonding. In steel, for example, a metallic bond predominates and the energy is readily transferred by the delocalized electrons. In polymers there are no delocalized electrons. Although energy can travel down the chain rather readily, it must eventually hop from one chain to another across relatively weak secondary bonds. This is much more difficult, and the conductivity of polymers is therefore relatively low.

Conduction in polymers is affected by a number of variables, including crystallinity, molecular weight, and the presence of fillers. Crystals increase thermal conductivity because the ordered lattice and close proximity of the

TABLE 7.2  Thermal Conductivities of Some Common
Materials (20°C)

| Material | Thermal Conductivity (W/m · K) |
|---|---|
| Copper | 401 |
| Aluminum | 237 |
| Iron | 80 |
| High-density polyethylene | 0.48 |
| Low-density polyethylene | 0.33 |
| Polytetrafluoroethylene | 0.25 |
| Nylon 6 | 0.24 |
| Polyformaldehyde | 0.23 |
| Epoxy | 0.19 |
| Polycarbonate | 0.20 |
| Polypropylene | 0.12 |
| Polystyrene | 0.11 |
| Still Air | 0.024 |

(*Adapted from* [1] *and* [3].)

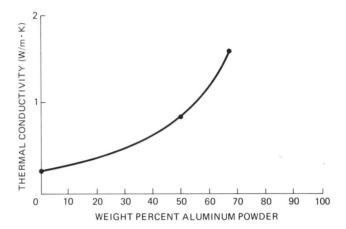

**Figure 7.2** Thermal Conductivity of Aluminum-Filled Epoxy. (*Adapted from* [4].)

molecules allow for better energy transfer. Alternatively, anything that disrupts this contact (such as melting or plasticizers) would be expected to reduce conductivity. Chain length and crosslinking are also factors, since longer chains require fewer energy jumps from one molecule to another.

The variations in conductivity that occur through changes in crystallinity, plasticizer content, and molecular weight are generally small compared to the differences between the conductivities of metals or ceramics and those of polymers. Thus it is not surprising that relatively large changes in the conductivity of polymers occur when fillers are added. Figure 7.2 illustrates an example of the conductivity changes that occur in epoxies when they are filled to various levels with aluminum powder. Alternatively, the conductivity of polymers can be reduced by incorporating air as a filler. Since still air has a very low conductivity, this effectively reduces the material cross section and gradient for heat transfer. The polystyrene foam panels used to insulate the walls and foundations of houses are one example of a commercial application of this principle.

## THERMAL EXPANSION

In most directions in most materials, an increase in temperature will result in an increase in the dimensions. Given a bar of length $l_0$ that is subjected to a one-degree rise in temperature, the linear thermal expansion coefficient, $\alpha$, is defined as

$$\alpha = \frac{\Delta l}{l_0}$$

where $\Delta l$ is the change in length associated with the unit temperature rise.

**TABLE 7.3  Linear Thermal Expansion Coefficients of Some
Common Materials (20–25°C)**

| Material | Coefficient of Linear Expansion ($10^{-6}$ m/m/K) |
|---|---|
| Glass | 8 |
| Iron | 12 |
| Copper | 17 |
| Aluminum | 25 |
| Epoxies | 45–65 |
| Polycarbonate | 68 |
| Phenolic | 68 |
| Polystyrene | 70–80 |
| Polypropylene | 81–100 |
| Nylon 6 (cast) | 90 |
| Polyformaldehyde | 100 |
| High-density polyethylene | 110–130 |
| Low-density polyethylene | 100–220 |

(*Adapted from* [1] *and* [3].)

The units for the coefficient of linear expansion are in/in/°F, cm/cm/°C or m/m/K, depending on whether the English, cgs, or SI system is used. Typical values for some common materials are given in Table 7.3. As indicated in this table, the expansion coefficients for polymeric materials are generally greater than those associated with common metals and ceramics. This is a direct result of the temperature sensitivity of the bonds that hold these atoms and molecules

**Figure 7.3**  Effect of Aluminum Powder on the Coefficient of Linear Expansion of a Cured Epoxy Resin. (*Adapted from* [4].)

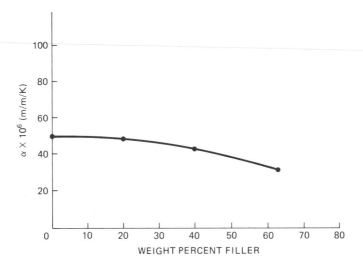

together to form the materials. In general, strong bonds are associated with relatively low coefficients of expansion and weak bonds with relatively high ones.

The importance of the relatively high thermal expansion coefficients of polymers cannot be overemphasized. A major problem can occur when a polymer is joined to a metal or a ceramic having a different thermal expansion coefficient. If the temperature of this composite system is changed in the course of production or use, significant stresses can occur in the polymer and at the interface. This effect is responsible in part for the failure of systems ranging from polymer overlays on concrete bridge decks to amalgam dental fillings in teeth.

Since the thermal expansion coefficients of metals and ceramics are typically lower than those of polymers, the expansion coefficients of polymers are lowered significantly when inorganic fillers are added. Figure 7.3 shows, for example, how the linear thermal expansion coefficient of an epoxy resin ($\alpha = 50 \times 10^{-6}$ m/m/K) is affected by the addition of aluminum powder ($\alpha = 0.25 \times 10^{-6}$ m/m/K).

## PROBLEM SET

1. Calculate the appropriate conversion factors (cgs, English, and SI systems) for:
   (a) Thermal conductivity
   (b) Specific heat
   (c) Thermal expansion

2. Using the equation that relates heat flux to thermal conductivity, show how the units for thermal conductivity are obtained.

3. In general, how does molecular weight influence thermal conductivity? Are the effects of changes in molecular weight the same in crystalline and amorphous polymers?

4. Why does the addition of plasticizer reduce the conductivity of PVC?

**Answer:**

Plasticizers disrupt the intermolecular bonds between the polymer molecules. This in turn makes it more difficult for energy to be transferred from one molecule to another. Thus thermal conductivity will decrease with increasing plasticizer content.

5. Would you expect atactic or isotactic polypropylene to have the higher thermal conductivity? Why?

6. Using the *Modern Plastics Encyclopedia*, compare the thermal conductivities of low- and high-density polyethylene. Explain any trend you find.

7. Calculate the energy necessary to raise the temperature of a cubic meter of steel and a cubic meter of polyethylene by one degree Kelvin.

8. Explain why plasticized forms of PVC have coefficients of linear expansion up to three times as large as for unplasticized PVC.

9. Since thermal expansion coefficients depend on bond stiffness and strength, how would you expect $\alpha$ to be affected by
   (a) Molecular orientation (such as occurs in fibers and films)?
   (b) Crosslinking?

**10.** Why is the thermal expansion of steel much lower than that of most polymers?

**11.** As noted previously, the stresses due to a mismatch in thermal expansion coefficients can be significant. As an example, consider a thin coating of foil on a thick slab of epoxy:

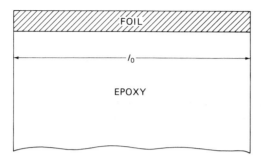

where

$$E_{foil} = 30 \times 10^6 \text{ psi} \qquad \alpha_{foil} = 10 \times 10^{-6}/K$$

$$E_{epoxy} = 0.5 \times 10^6 \text{ psi} \qquad \alpha_{epoxy} = 400 \times 10^{-6}/K$$

If this system is subjected to a change in temperature, $\Delta T$, of 200°C, what is the resulting tensile stress in the foil?

**12.** Cite several examples that illustrate the importance of specific heat, thermal expansion (contraction), and thermal conductivity in:
**(a)** The fabrication of polymeric products
**(b)** The use of polymeric products

## REFERENCES

[1] Chemical Rubber Company, *Handbook of Chemistry and Physics*, 59th ed. (R. C. Weast, ed.), CRC Press, West Palm Beach, Fl. (1978–1979).

[2] Wilski, H., Heat Capacity of High Polymers, in *Polymer Handbook*, 2nd ed. (J. Brandrup and E. H. Immergut, eds.), Wiley, New York (1975).

[3] *Modern Plastics Encyclopedia*, McGraw-Hill, New York (1980–1981).

[4] Lee, H., and Neville, *Handbook of Epoxy Resins*, McGraw-Hill, New York (1967).

The temperatures at which polymers melt and soften are important in their fabrication and use, just as melt temperatures and softening temperatures are important in the production and use of other crystalline solids and glasses. In polymers the molecules can exist in either an amorphous or a crystalline state, so two different types of thermal transitions must be considered. The melt transition is associated with the crystalline regions of polymers, and the temperature at which it occurs is $T_m$. The glass transition is associated with amorphous polymers and amorphous regions of crystalline polymers, and the temperature at which it occurs is $T_g$. Both of these transitions and some of the characteristics of polymers that affect them are considered in this chapter.

## CRYSTALLINE MELT TRANSITION

Everyone at one time or another has probably watched ice, wax, or other crystalline materials melt and is familiar with the types of changes that can occur during melting. Melting is a process in which an ordered, crystalline structure is replaced by a random, fluid one. It is not a degradative process, but rather a rearrangement of the molecules (in the case of a molecular solid) or atoms (in the case of an atomic solid). Since crystals are important determinants of many properties, large changes in modulus, strength, hardness, conductivity, etc. can occur when they melt. Therefore melt temperatures can have much bearing on the use and fabrication temperatures of materials.

Most metals melt at relatively high temperatures. For example, pure

TABLE 8.1  Crystal Melt Temperatures
of Common Polymers

| Polymer | $T_m (°C)$ |
|---|---|
| Polyethylene (low-density) | 95–130 |
| Polyethylene (high-density) | 120–140 |
| Polypropylene | 168 |
| Polyformaldehyde | 181 |
| Nylon 6 | 216 |
| Polyethylene terephthalate | 245 |
| Nylon 6/6 | 265 |
| Polytetrafluoroethylene | 327 |

(*Adapted from* [2].)

aluminum melts at approximately 660° C, pure copper at approximately 1083° C, and pure iron at approximately 1535°C [1]. Polymers melt at much lower temperatures than these metals; with few exceptions the crystal melt temperatures of polymers are below 300°C. Some of these values are listed in Table 8.1. It is readily evident from this table that, although the $T_m$'s are lower than those of most metals, some polymers melt at much higher temperatures than others. For example, the polyethylenes have relatively low melt temperatures, ranging from 95°C for highly branched material to approximately 140°C for linear material. Teflon* (polytetrafluoroethylene), on the other hand, has a melt temperature of 327°C.

The Gibbs free energy is a useful tool for understanding why some polymers melt at higher temperatures than others [3]. For a constant-temperature process the Gibbs free energy change, $\Delta G$, is given by

$$\Delta G = \Delta H - T\Delta S$$

where $\Delta H$ is the change in enthalpy, $T$ is the absolute temperature, and $\Delta S$ is the change in entropy. For the crystal melt transition this equation becomes

$$\Delta G_m = \Delta H_m - T_m \Delta S_m$$

where $T_m$ is the melt temperature and the quantity of interest. The melting of crystals is a reversible process, since a small change in the process conditions will shift it in the opposite direction; consequently the change in the Gibbs free energy is zero for melting. By substituting $\Delta G_m = 0$ into the previous equation, one finds that

$$T_m = \frac{\Delta H_m}{\Delta S_m}$$

The temperature at which a polymer melts therefore depends on the relationship between these two thermodynamic quantities.

*Teflon is a du Pont trademark.

The change in enthalpy, $\Delta H_m$, is related to the strength of the bonds that hold the crystal together. A crystal that is strongly bonded requires more energy to melt than one that is weakly bonded. For example, the enthalpy change tends to be large when hydrogen-bonded crystals melt and lower when van der Waals-bonded crystals melt. An example of this is the nylons (polyamides). Although all

nylons contain the $-\overset{\overset{\displaystyle H}{|}}{N}-\overset{\overset{\displaystyle O}{\|}}{C}-$ group, they differ in regard to the other components of the backbone of the chain. For example, nylon 6/6 has the following repeating unit:

$$\begin{array}{c} \Big[ \overset{\overset{\displaystyle H}{|}}{N}-(CH_2)_6-\overset{\overset{\displaystyle H}{|}}{N}-\overset{\overset{\displaystyle O}{\|}}{C}-(CH_2)_4-\overset{\overset{\displaystyle O}{\|}}{C} \Big]_n \end{array}$$

while the repeating unit of nylon 6/10 is

$$\begin{array}{c} H-\Big[ \overset{\overset{\displaystyle H}{|}}{N}-(CH_2)_6-\overset{\overset{\displaystyle H}{|}}{N}-\overset{\overset{\displaystyle O}{\|}}{C}-(CH_2)_8-\overset{\overset{\displaystyle O}{\|}}{C} \Big]_n \end{array}$$

The melt temperature of nylon 6/6 (265°C) is much higher than that of nylon 6/10 (215°C) because the distance between the hydrogen bonding sites (in terms of $-CH_2-$ links) is greater in nylon 6/10. This in turn results in weaker intermolecular bonds and a smaller $\Delta H_m$ for nylon 6/10.

The change in entropy, $\Delta S_m$, is the increase in disorder or randomness that occurs during melting. It is a positive change, since the crystal is an ordered state whereas the melt or fluid is a disordered one. A complication with polymers is that in the melt, varying degrees of randomness that affect $T_m$ can exist. This degree of randomness depends on both the molecular weight and the stiffness of the polymer molecules.

The molecular weight dependence of $\Delta S_m$ can be visualized in the following manner. Assume that three situations exist, each involving 1000 monomer units. In the first case these units are linked together to form one molecule having a degree of polymerization equal to 1000; in the second case the units are divided among 10 molecules; and in the third case the units are 1000 individual molecules. These cases are illustrated in Figure 8.1. The first case has the lowest entropy of the three, because each of the 1000 units is more restricted in the number of possible positions that it can have than in the other two cases. In the first case each unit must remain attached to its neighbors, whereas in the third case this restriction does not apply. Thus, increasing the molecular weight of a polymer reduces the entropy change that occurs during melting. Changes in molecular weight generally have larger effects on $T_m$ at low average molecular weights than at high average molecular weights. This effect can be seen by comparing the melt temperature ranges of hexane (−95°C), paraffin wax (50–60°C), high-density polyethylene (120–140°C), and ultrahigh-molecular-weight polyethylene (125–140°C).

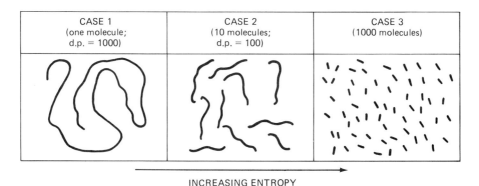

| CASE 1 (one molecule; d.p. = 1000) | CASE 2 (10 molecules; d.p. = 100) | CASE 3 (1000 molecules) |

INCREASING ENTROPY

**Figure 8.1** Effect of Chain Length on the Randomness of the Melt.

The entropy change during melting is also influenced by the stiffness or flexibility of the molecules. Many parameters influence this stiffness, and these will be discussed in greater detail in a later section. For the present purpose it is sufficient to note that a very stiff molecule can assume fewer conformations than a very flexible one. The entropy of the melt of a stiff molecule will therefore be lower than that of a more flexible one, and the total $\Delta S_m$ will be smaller for the stiffer molecules.

A classic illustration of the influence of entropy change on $T_m$ is polytetrafluoroethylene:

$$\left[\begin{array}{c} F\ \ F \\ | \ \ | \\ C-C \\ | \ \ | \\ F\ \ F \end{array}\right]_n$$

Its melt temperature is extremely high for a polymer (327°C). At first one might be tempted to attribute the high $T_m$ to the dipolar nature of the carbon–fluorine bonds; but on close inspection the fallacy of this approach becomes evident. A set of molecular models that accurately depict the sizes of the atomic components would show that the electronegative fluorine atoms shield the electropositive carbon backbone, making a significant positive–negative interchain attraction unlikely. If the high melt temperature is not attributable to a large enthalpy change, then it must involve a very small entropy change. In polytetrafluoroethylene the four electronegative fluorine atoms on the backbone carbons make rotation about the carbon–carbon atoms very difficult. For reasons that will be explained later, this means that polytetrafluoroethylene molecules are very rigid compared to other molecules such as polyethylene. In addition, a typical molecular weight might be $3 \times 10^6$ g/mol, which corresponds to a degree of polymerization of 30,000. This is large even by polymeric standards. This combination of a stiff molecule and a long molecule results in an extremely low $\Delta S_m$ and a correspondingly high $T_m$ for polytetrafluoroethylene.

## THE GLASS TRANSITION

The crystal melt temperature involves the crystalline regions of a polymer. In polymers that have no crystals and in the amorphous regions of crystalline polymers, a different softening or melting process called the *glass transition* occurs. Anyone who has watched a glass blower at work has seen what happens at a glass transition. Glass is transparent and has no crystals, but there is a temperature above which the material softens. This change occurs more gradually than a melting transition, but it is still rather abrupt, as shown in Figure 8.2. This also happens at the glass transitions of polymers, except at much lower temperatures. Atactic polystyrene, for example, is a rigid, brittle, amorphous (glassy) polymer at room temperature. Its modulus versus temperature curve is similar to that of the inorganic glass illustrated below, except that its glass transition temperature ($T_g$) is approximately 100°C. To fabricate a product from polystyrene, therefore, one must approach or exceed $T_g$ only; there are no crystals to be melted in this case.

In a similar fashion, a piece of flexible rubber tubing immersed in liquid nitrogen (77 K) for a short time will become brittle and can be shattered, like a piece of glass. This occurs because the rubber at room temperature is somewhat like the inorganic glass above its softening temperature. It is completely amorphous and would be a fluid (like the glass) were it not for the crosslinks (see Chapter 14). In cooling to 77 K its stiffness changes, as shown in Figure 8.3, with most of the change occurring in a relatively narrow region around $T_g$. Similar effects are noted in the amorphous regions of semicrystalline polymers (Figure 8.4).

In a practical sense, therefore, the glass transition temperature of a polymer can be defined as the temperature below which an amorphous polymer (or an amorphous region of a crystalline polymer) is rigid and brittle (glassy) and above which it is rubbery or fluidlike. Table 8.2 lists the $T_g$'s of some common polymers. As was the case with $T_m$, the glass transition temperatures of various polymers cover a wide range. Silastic rubber, for example, has a relatively low $T_g$ of

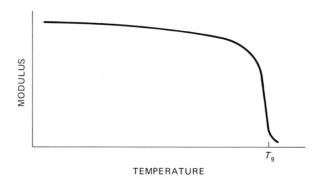

**Figure 8.2** Modulus versus Temperature for an Amorphous Material.

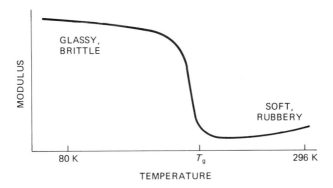

**Figure 8.3** Modulus versus Temperature for a Crosslinked Polymer That is Rubbery at Room Temperature.

$-123°$C, whereas polycarbonate has a relatively high $T_g$ of $150°$C. The location of $T_g$ relative to the application temperature(s) of the product can be very important, since it has a strong influence on properties such as modulus. Rubbery materials require $T_g$'s below their application temperatures (Chapter 14), whereas other amorphous systems, such as polycarbonate, polyether sulfone, and polystyrene are useful only at temperatures below $T_g$. Semicrystalline polymers are utilized either above or below their $T_g$'s, depending on the polymer and the application.

Although the nature of the glass transition in polymers is not completely understood, there appears to be a correlation between the temperature at which the glass transition occurs for a given polymer and the flexibility of the molecules. In general, the glass transition temperature tends to be lower for a polymer consisting of inherently flexible molecules than for those consisting of inherently stiff molecules. This is reasonable if one assumes that a direct relationship exists between molecular flexibility and the modulus of the bulk polymer; that is, the molecules are more rigid below $T_g$ than they are above $T_g$.

Many parameters can influence the stiffness or flexibility of polymer

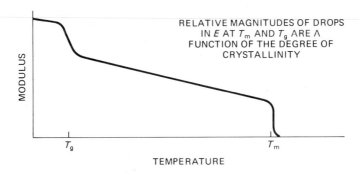

**Figure 8.4** Modulus versus Temperature for a Semicrystalline Polymer.

TABLE 8.2  Glass Transition Temperatures of Some Common
Polymers as Reported by Various Investigators

| Polymer | $T_g(°C)$ |
|---|---|
| Polydimethyl siloxane | −123 |
| Polyethylene | −120 |
| Polyisoprene | −73 |
| Polyoxymethylene (polyformaldehyde) | −50, −85* |
| Polypropylene | −10, −18* |
| Nylon 6/10 | 40 |
| Nylon 6, 6/6 | 50 |
| Polyethylene terephthalate | 69 |
| Polyvinyl chloride | 87 |
| Polystyrene | 100 |
| Polyacrylonitrile | 104, 130* |
| Epoxies | generally > 100 |
| Polytetrafluoroethylene | 126 |
| Polycarbonate | 150 |
| Polydiphenylether sulfone | 230 |

*More than one value reported.
(*Adapted from* [2], [4], *and* [5].)

molecules. Interactions between a given molecule and its neighbors are important, as are the inherent molecular structural characteristics. Some of these characteristics are the following:

1. The elements that compose the backbone of the polymer (e.g., carbon-carbon single and double bonds, phenylene groups, oxygen linkages)
2. The size and position of side groups
3. Intermolecular bond strength
4. Crosslinking
5. Molecular weight

Several of these are examined in detail in the sections that follow, both to indicate why differences in $T_g$ values occur and also to illustrate how these characteristics can be altered to suit specific applications.

### Side Groups

In a polymer whose backbone consists of carbon atoms, the presence of atoms or molecular groups different from hydrogen on the other two bonds has a great influence on the flexibility of the molecule, since molecular motion in this case involves rotation about the carbon–carbon single bonds. That is, in order for the molecular segment

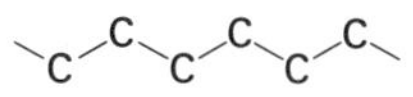

to move to the position

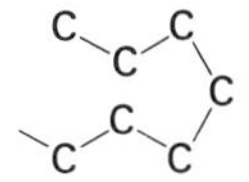

some rotation about the single bonds (rather than bending of those bonds) must occur:

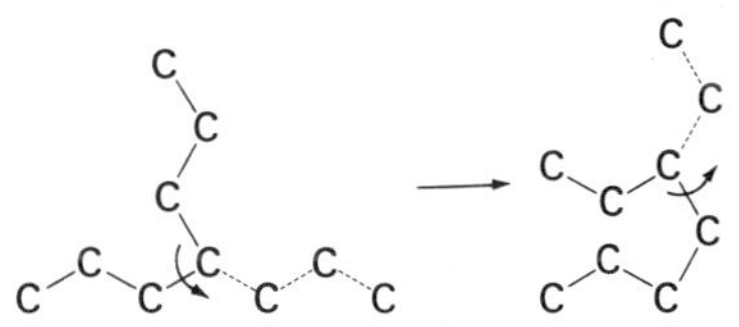

The degree to which the molecule can rotate about these bonds depends to some extent on the side groups and the amount of space they require. Consider, for example, the following chain segment:

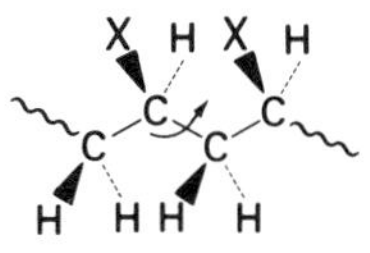

where X can be any of the typical side groups that occur in polymers. If X is a hydrogen atom (the smallest unit possible), then the interference it causes with its neighbors (both on the chain and on neighboring chains) in rotating around the central carbon bond is minimal. This is the case in polyethylene, which has a relatively low $T_g$. If X is large (such as the phenyl group on polystyrene) or charged (such as the chlorine atom on polyvinyl chloride), then the interference will be much larger. This stiffens the molecule and raises the $T_g$.

An interesting application of this principle is in polyvinyl chloride-related pipe products. One can purchase both PVC pipe and CPVC pipe, the latter being more suitable for higher-temperature (e.g., domestic hot water) applications. CPVC is a chlorinated version of PVC; that is, some additional chlorine atoms, rather than hydrogen atoms, occupy the side group positions of the molecules. This increases the $T_g$ of the material because of the additional rigidity imparted to the molecules by the chlorine atoms. Without this chlorination PVC pipe is unsuitable for conveying hot liquids because the high temperature of the liquids causes PVC to soften. Its $T_g$ is too close to the liquid temperature.

### Intermolecular Bond Strength

Since molecules do not move independently of one another, the secondary bonds between molecules are important. When these interactions are strong, the molecules have more difficulty moving and the $T_g$ tends to be higher. The intermolecular bond strength can be deliberately reduced by the use of plasticizers. A plasticizer is a compound of relatively low molecular weight that is added to some polymers during processing. Plasticizers reduce intermolecular bond strength by positioning themselves between the polymer molecules, thus separating them and lessening their interactions. This allows the polymer molecules to move more readily, and the glass transition temperature is lowered when this occurs.

One of the earliest uses of synthetic plasticizers was the addition of camphor to cellulose nitrates. Today the most important use of plasticizers is in polyvinyl chloride products. In a largely unplasticized form, PVC is used in applications such as pipe and siding, which require a relatively high $T_g$. Many of the other uses for PVC, such as seat covers, floor tile, tablecloths, and shower curtains require a more flexible material at room temperature. This flexibility is obtained by adding plasticizers to PVC, which lower the $T_g$ below the use temperature of the item (see Figure 8.5). For example, a flexible vinyl garden hose is a plasticized form of PVC. On a warm day it is soft and flexible, but on a cold day it becomes rather stiff and difficult to handle because the ambient temperature is approaching its glass transition temperature. Plasticizers are covered in greater detail in the chapter on solution properties.

### Crosslinking

Crosslinking also has a significant effect on the glass transition temperature. A crosslink tightly bonds two molecular segments together and restricts their motion. As the crosslinking density increases, the restrictions become more

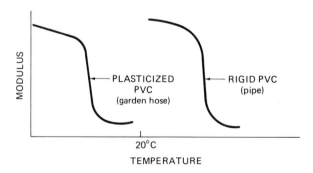

**Figure 8.5** Modulus versus Temperature for Plasticized and Unplasticized PVC.

widespread in the structure and the $T_g$ increases. Very heavily crosslinked polymers such as urea formaldehyde, phenol formaldehyde, and some epoxies have relatively high glass transition temperatures, and often degrade below their $T_g$'s.

An example of the deliberate modification of crosslinking density to obtain specific properties is in polyurethane foams. These foams are produced by the reaction of an isocyanate and a polyol (see Chapter 16). A blowing agent (either produced during the reaction or added separately) is used to expand the polymerizing mass and create an air–polymer mixture having a relatively low density. Polyurethane foams are either flexible or rigid at room temperature. These are very common products; flexible foams are used primarily for upholstery applications, and rigid foams are frequently used for thermal insulation. Structurally, the major difference between a flexible urethane foam and a rigid one is the degree of crosslinking. Flexible foams are lightly crosslinked, and their $T_g$'s are below room temperature. Rigid foams have higher degrees of crosslinking, and their $T_g$'s are above room temperature.

## PROBLEM SET

1. Compare the general shapes of modulus versus temperature curves for the following:
   (a) A completely amorphous, uncrosslinked polymer
   (b) A semicrystalline polymer with high crystallinity
   (c) A semicrystalline polymer with low crystallinity
   Be sure to label the axes and indicate pertinent transitions.

2. On the same set of axes, draw approximate modulus versus temperature curves for:
   (a) Polyethylene
   (b) Nylon
   (c) Iron
   (d) Aluminum
   What does this tell you about the comparative working ranges and processing temperatures of these materials?

3. Which of the polymers in Table 3.2 are amorphous and which are semicrystalline?

4. For which of the polymers in Table 3.2 is room temperature:
   (a) Above $T_g$?
   (b) Below $T_g$?
   (c) Below $T_g$ and $T_m$ (where applicable)?
   (d) Between $T_g$ and $T_m$ (where applicable)?

5. Explain:
   (a) How molecular weight influences $\Delta S_m$
   (b) How molecular flexibility influences $\Delta S_m$

6. Indicate which member of the following pairs has the *lower* melt temperature. Briefly explain why in terms of $\Delta H_m$ and $\Delta S_m$.
   (a) A polymer with crystals containing few defects or one containing crystals with many defects (assume that they have the same repeating unit)
   (b) A polymer such as nylon in which hydrogen bonds hold the crystals together or a

polymer such as polypropylene in which van der Waals forces hold the crystals together

(c) Polyethylene or polypropylene

7. PET (polyethylene terephthalate) and PBT (polybutylene terephthalate) are both thermoplastic polyester resins. The $T_m$ of PBT is approximately 40° C lower than that of PET. Explain why this difference exists, in terms of $\Delta H_m$ and $\Delta S_m$. The repeating units are

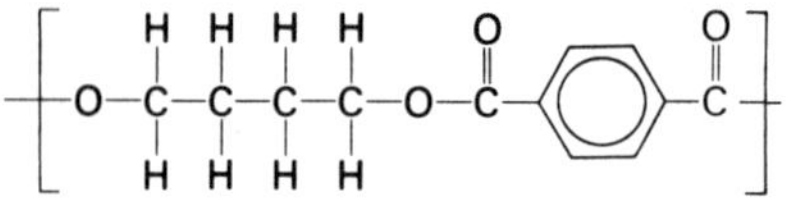

PBT 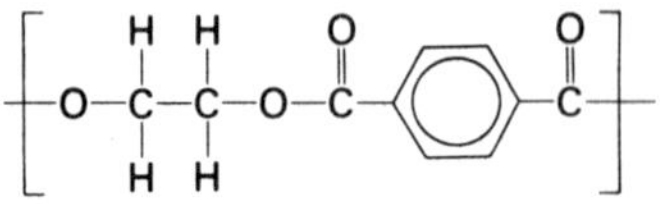                                PET

8. How does the magnitude of $T_g$ compare with that of $T_m$ for a given semicrystalline polymer (use Tables 8.1 and 8.2 for comparisons)?

9. What is a plasticizer? How does it affect $T_g$?

10. Compare the glass transition temperatures of polyethylene, polypropylene, and polystyrene. How do these temperatures relate to the size of the side groups?

11. Explain the rationale behind the use of CPVC in pipe and pipe fittings.

12. Molecular flexibility is also related to the molecular groups that form the backbone of the molecule. How do the following compare with carbon–carbon single bonds in regard to flexibility?

(a)

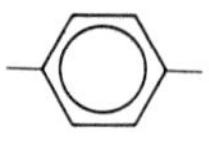

(b)   $-O-$

(c)

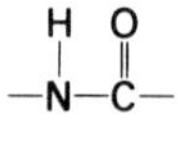

(d)   $C=C$

13. Aromatic thermoplastic polyimides are amorphous polymers with very high $T_g$'s (over 300° C). Their structures are of the form

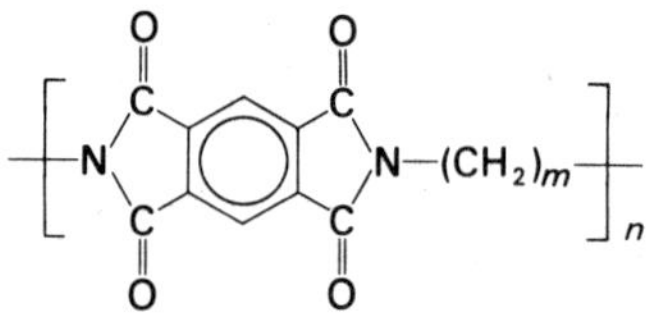

In light of your answer in Problem 12, what is one reason for their high $T_g$'s?

## REFERENCES

[1] Chemical Rubber Company, *Handbook of Chemistry and Physics*, 59th ed. (R. C. Weast, ed.), CRC Press, West Palm Beach, Fl. (1978–1979).

[2] *Modern Plastics Encyclopedia*, McGraw-Hill, New York (1980–1981).

[3] Rosen, S. L., *Fundamental Principles of Polymeric Materials*, Wiley, New York (1981).

[4] Nielsen, L. E., *Mechanical Properties of Polymers and Composites*, vols. 1 and 2, Dekker, New York (1974).

[5] Allcock, H. R., and F. W. Lampe, *Contemporary Polymer Chemistry*, Prentice-Hall, Englewood Cliffs, N.J. (1981).

A polymer's molecular structure is not necessarily constant during the lifetime of a product. Instead, the molecules and the resulting bulk properties of the material may permanently change with time because of various degradative elements in their environment. Frequently, for instance, a vacuum hose in an engine compartment becomes brittle after a few years of service, or a polyethylene film cracks after a few months' exposure to sunlight, or a tire develops surface cracking after a few thousand miles of travel. These are all examples of different types of degradation that can occur in polymers.

In general, these changes occur as a result of specific and combined effects of heat, light, and chemical reactants (such as oxygen and ozone) on the structures of polymers. All polymers are not equally susceptible to the effects of these agents. A few of the more common cases are discussed below as examples.

## OXIDATIVE DEGRADATION OF POLYOLEFINS

In polyethylene and polypropylene significant degradation can occur in the presence of oxygen when these materials are exposed to heat or ultraviolet radiation. The susceptible sites are the hydrogen atoms that share a carbon atom with either a branch on the polyethylene molecule:

```
                              H
                              |
                          H—C—H
                              |
                          H—C—H
            H                 |
            |             H—C—H
        H—C—H                 |
            |             H—C—H
        H—C—H                 |
            |             H—C—H
      ~~~C~~~~~~~~~~~~~~~C~~~
         (H)               (H)
            \           /
```

or a methyl group on the polypropylene repeating unit:

```
        H  CH₃
        |   |
    ~~~C—C~~~
        |   |
        H  (H)
```

Since polypropylene has more of these tertiary hydrogens, it is more sensitive than polyethylene to this type of degradation.

The chemical process that occurs is complex. It typically involves the generation of free radicals, which react with oxygen and eventually lead to processes such as scission of the polymer chains. This decreases the molecular weight, and the properties decline accordingly. As the process progresses, some crosslinking also occurs, which makes the material insoluble and brittle. A description of the chemical reactions that take place can be obtained in Saunders [1] and in other polymer texts [2, 3, 4]. The most important point is that the process permanently alters the structure of the molecules and can result in premature failure of the product.

This type of degradation can be retarded with antioxidants.* These additives are generally aromatic amines or phenols, which contain relatively unstable hydrogen atoms. Their function is to deactivate the free radicals that form during the degradative process and thereby prevent the scission and crosslinking reactions. Another approach to the problem is the addition of other additives (such as carbon black) that absorb ultraviolet radiation. Black polyethylene film, for example, typically retains its useful properties outdoors much longer than clear polyethylene largely because of this absorption.

---

*Antioxidants are also used in foods to prevent them from becoming stale. A common one is butylated hydroxytoluol (BHT).

## DEGRADATION OF POLYVINYL CHLORIDE

When PVC is exposed to heat or ultraviolet radiation, it will darken, its weight will decrease, and it will become brittle and insoluble. Two separate processes are believed to be involved: One is an oxidative process similar to that in the previous discussion, and the other is dehydrochlorination. The latter process involves the removal of hydrogen chloride from the polymer chains:

$$
\begin{array}{c}
\phantom{x}(H \quad Cl)(H \quad Cl) \\
\phantom{x}|\phantom{xx}|\phantom{xxx}|\phantom{xx}| \\
\sim\!\!\sim C-C-C-C\sim\!\!\sim \\
\phantom{x}|\phantom{xx}|\phantom{xx}|\phantom{xx}| \\
\phantom{x}H \quad H \quad H \quad H
\end{array}
$$

and the formation of a series of conjugated double bonds:

$$
\begin{array}{c}
\sim\!\!\sim C=C-C=C\sim\!\!\sim \\
\phantom{x}|\phantom{xx}|\phantom{xx}|\phantom{xx}| \\
\phantom{x}H \quad H \quad H \quad H
\end{array}
$$

The HCl tends to be removed in a progressive (rather than random) fashion from the chain. This is called an unzipping reaction, and it is assumed to start at chain ends, branch points, and sites of previous oxidative degradation. The resulting arrangement of double bonds (above) is a color-sensitive group, and it is the cause of the observed darkening.

Various stabilizers are used to prevent this type of degradation in PVC. They include basic lead carbonate, dibasic lead phthalate, barium laurate, and epoxidized linseed oil. The mechanisms by which these compounds work are not well known, but some details are available in Saunders [1] and Rodriguez [2].

## OZONE DEGRADATION OF DIENES

Natural rubber and other unsaturated* elastomers (such as styrene butadiene rubber) are particularly susceptible to ozone degradation. Ozone ($O_3$) is thought to attack the carbon–carbon double bonds in these polymers and eventually divide the molecule. This can be a significant problem, particularly in high-voltage applications where ozone concentrations tend to be high. Ozone degradation can be retarded with waxes (ozonates) that exude to the surface of the polymer and therefore retard the diffusion of ozone into the polymer. Other products that react directly with ozone can also be used.

---

*The term "unsaturated" refers to the presence of carbon–carbon double bonds. A saturated material is one that lacks these bonds. This terminology is identical to that used in the controversies about saturated versus unsaturated fats and heart disease.

## HEAT AGING

Although these three examples deal with specific degradative mechanisms, all polymers are affected to some extent by long-term exposure to elevated temperatures. This phenomenon is called heat aging, and many processes (including those just discussed) may be involved, depending on the structure of the polymer and any additives that are present.

Many applications for polymers involve exposure to elevated temperature. These include automobile engine components, electrical and electronic devices in appliances, clothing and textiles, and sterilizable biomedical products. Thus it is very important in these and similar applications to understand the temperature-exposure limits of polymers being considered for the product. A knowledge of $T_m$ and $T_g$ and the limits they place on the application may not be sufficient because of these degradative effects, which can occur in the long term. For applicatons in which the mechanical stresses are not large (such as many electrical devices), the upper use temperature of a polymer can be estimated by the Underwriters' Laboratory (U.L.) temperature index. Typical values of this index for some polymers are listed in Table 9.1.

This index is obtained by exposing samples to elevated temperatures for various times and monitoring the decline that occurs in specific mechanical and electrical properties. The times required at various temperatures to reduce the original property by a factor of one half are determined. These values are then used to plot logarithm of time versus the reciprocal of absolute temperature (see Figure 9.1). The temperature index is the temperature at which this curve predicts failure at a specified time (usually $1.1 \times 10^4$ hours). This is a function of both the property measured and the thickness of the material.

The U.L. temperature index is useful as a rough guide for use at elevated temperatures, but actual in-service performance may depend on additional

TABLE 9.1 U.L. Temperature Indices
of Various Generic Polymers

| Polymer | Temperature Index |
|---|---|
| Nylon 6, 6/6, 6/10 | 65°C |
| Polycarbonate | 65°C |
| Silicone rubber | 105°C |
| Epoxy molding compounds* | 130°C |
| Thermosetting polyester molding compounds* | 130°C |
| Phenolic molding compounds* | 150°C |

*Not including organic-fiber-filled systems or filled liquid resin systems.

(*Adapted from* [5].)

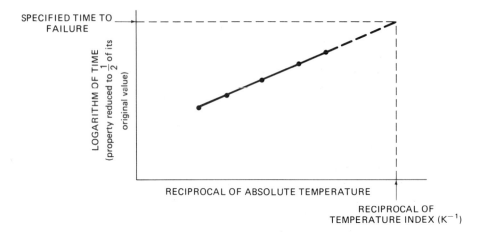

**Figure 9.1** Determination of the U.L. Temperature Index. (*Adapted from* [5].)

variables in the material's environment, which are not considered in this test. Ultraviolet radiation, ozone, airborne contaminants, and solvents may be important in specific applications. In addition, mechanical stress will influence aging characteristics. These are not variables in the U.L. temperature index determination, but they should be considered where applicable. When a product is under considerable mechanical stress, creep rupture (Chapter 13) may provide a better estimate of a material's aging characteristics at elevated temperatures than the U.L. temperature index [5].

## PROBLEM SET

1. Explain why polyethylene might be chosen over polypropylene for an exterior application.
2. Why are many polymers used in exterior applications black? What is one disadvantage of this?
3. If PVC becomes brittle and loses weight, does that necessarily mean that it has degraded? Explain.
4. Explain the relationship between suntans and polymer degradation. How do suntan oils and ointments work? (Suggested reference: [2].)
5. Why is ethylene–propylene rubber (see Chapter 14) more resistant to ozone attack than styrene–butadiene rubber?
6. The U.L. temperature index is a function of the type of test used. Obtain a recent copy of *Modern Plastics Encyclopedia* and answer the following:
   (a) What are the general types of tests used?
   (b) Which combination of tests appears to be the most severe?
   (c) In general, how large are the index differences among the test combinations?

7. Temperature index can vary a great deal for a given type of polymer. Using the *Modern Plastics Encyclopedia*, examine the range of values for polyphenylene oxide-based resins.

## REFERENCES

[1] Saunders, K. J., *Organic Polymer Chemistry*, Chapman and Hall, London (1973).

[2] Rodriguez, F., *Principles of Polymer Systems*, McGraw-Hill, New York (1970).

[3] Allcock, H. R., and R. W. Lampe, *Contemporary Polymer Chemistry*, Prentice-Hall, Englewood Cliffs, N.J. (1981).

[4] Seymour, R. B., and C. E. Carraher, Jr., *Polymer Chemistry*, Dekker, New York (1981).

[5] *Modern Plastics Encyclopedia*, McGraw-Hill, New York (1980–1981).

In the United States approximately 12,000 people die each year because of fire* [1]. In a majority of these cases the ignition of natural and/or synthetic polymers is involved. Because of this there is growing concern regarding the use of polymers in various markets such as textiles, transportation, electronic products, and housing. This has resulted in the development of a wide variety of flame-retardant polymeric materials. The following is an introduction to this group of materials and to the general field of fire retardancy.

## THE MEASUREMENT OF FLAMMABILITY

No single test or series of different tests can accurately assess the fire hazard which a material or product will pose in all situations. However, there are a number of properties that are involved to varying degrees in fire safety and are the basis for the regulation of fire-retardant products. These properties are the following:

1. The ease with which a material can be ignited, or, alternately, the ease with which an ignited material can be extinguished.
2. The rate at which a flame travels across the surface of a material or product.
3. The rate at which a flame can penetrate a material or product (as opposed to traveling across its surface).

*In comparison, approximately 49,500 people died in automobile accidents in the United States in 1977.

4. The total BTU content of the material.

5. The rate at which heat is released as the material burns.

6. Smoke evolution.

7. Toxic gas evolution.

For a specific application, some of these factors are more important than others. For example, in the regulation of children's sleepwear, total BTU content is of much less importance than ignitability and rate of surface flamespread. On the other hand, BTU content is critical in assessing the fuel load in a building. Some of the tests that measure these properties are listed in Table 10.1.

Existing tests represent the standards by which fire-retardant polymers are developed and tested. One important and widely used test, which will serve as an example, is ASTM D 2863 [2]. This is a relatively simple test that measures the ease of extinction of a burning material. A schematic diagram of the apparatus is shown in Figure 10.1. The major component is a glass column, which houses the burning sample. The relative flow rates of nitrogen and oxygen through the combustion column are adjusted to obtain the minimum concentration of oxygen that will allow the sample to support a flame. This minimum value is the limiting oxygen index, $n$, and it is calculated from

$$n = \frac{100(O_2)}{(O_2 + N_2)}$$

where $O_2$ and $N_2$ are the volumetric flow rates of oxygen and nitrogen,

TABLE 10-1  Some Common Tests for the Flammability
of Polymers and Polymeric Products

| Test | Comments | Reference |
|---|---|---|
| Limiting oxygen index (ASTM D 2863) | See text | [2] |
| Steiner tunnel test (ASTM E 84) | Measures smoke, flamespread, and fuel contributions of various building materials | [3] |
| ASTM E 119 | Measures rate of fire penetration for wall sections, coverings, etc. | [3] |
| NBS smoke chamber (ASTM E 662) | Measures smoke emissions from various materials | [3] |
| ASTM D 635 | Measures burn rate of a variety of self-supporting polymers | [2] |
| ASTM D 1929 | Measures ignition temperatures of polymers using a hot-air ignition furnace | [2] |
| Radiant panel test (ASTM E 162) | Measures flamespread using a radiant-heat energy source | [2] |
| ASTM D 1692 | Measures burn rate of rigid or flexible foams | [2] |

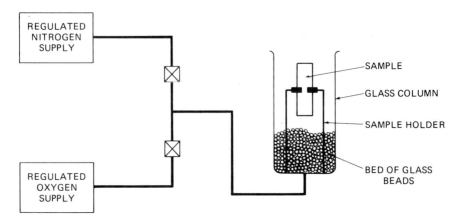

**Figure 10.1** Limiting Oxygen Index Test.

respectively. The reference value is 0.21, which is the approximate percentage of oxygen in the atmosphere. A material with a limiting oxygen index less than this value will support a well-defined flame in air, but one with a higher limiting oxygen index will not. Limiting oxygen indices for a few polymers are given in Table 10.2

As is evident from this short list, synthetic polymers exhibit a wide range of values. For example, unplasticized polyvinyl chloride has a relatively high oxygen index of 0.45, while polymethyl methacrylate has a relatively low oxygen index of 0.173. Similarly, plasticized polyvinyl chloride has a significantly lower oxygen index than unplasticized PVC. These differences derive to a large extent from differences in the structures and compositions of these materials. To understand why some of these differences exist, it is necessary to understand first

**TABLE 10.2  Limiting Oxygen Indices for Some Typical Polymers**

| Polymer | Limiting Oxygen Index |
|---|---|
| Polyformaldehyde | 0.15 |
| Polymethyl methacrylate | 0.173 |
| Polypropylene | 0.175–0.18 |
| Polyethylene | 0.174 |
| Polystyrene | 0.183 |
| Polycarbonate | 0.26–0.28 |
| Polyphenylene oxide | 0.285 |
| Nylon 6 | 0.23–0.26 |
| Nylon 6/6 | 0.24–0.26 |
| Silicone rubber | 0.26–0.34 |
| Polyvinyl chloride (unplasticized) | 0.45 |
| (plasticized) | <0.45 |
| Polytetrafluoroethylene | 0.95 |

(*Adapted from* [4] *and* [5].)

the combustion process and the basic fire-retardant mechanisms. These are discussed in the following sections.

## THE COMBUSTION PROCESS

The combustion of solids is an extremely complex process, but for the purposes of this text it can be visualized as occurring in a series of six zones. Three of these zones exist in the burning material, and three exist in the surrounding environment. Figure 10.2 is a schematic representation of these zones.

The three zones within the burning material are the intact region, the heating zone, and the degradation zone. The intact region is that portion of the material in which no effects of the fire are evident. The size and location of this region depends on the thermal conductivity and the burning time. The heating zone exists above the intact region and is where the molecules are beginning to react to the heat generated by combustion. The absorption of this thermal energy increases the temperature of the molecules (perhaps raising them above $T_g$ or

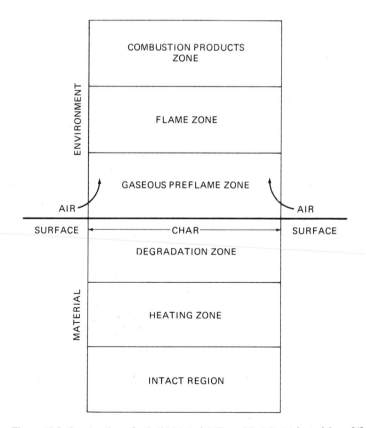

**Figure 10.2** Combustion of a Solid Material (Zone Model). (*Adapted from* [6].)

$T_m$), and, in the case of hygroscopic substances such as nylon, wool, wood, and cotton, can cause substantial dehydration. When these molecules absorb sufficient energy, some of their primary bonds begin to break and the polymer is reduced to various compounds of lower molecular weight. These compounds can be either solid (char), liquid (tar*), or gas. This breakdown occurs in the degradation (pyrolysis) zone, which exists at and below the surface of the polymer.

The three zones above the surface of the burning material are the gaseous preflame zone, the flame zone, and the combustion products zone. The gaseous preflame zone is a region directly above the surface of the polymer. It is in this region that the gases and tars produced in the degradation zone mix with oxygen from the surroundings. Eventually, sufficient oxygen mixes with these gases and tars that they ignite, producing large quantities of energy in the form of heat and light. This occurs in the flame zone. Much of the heat produced here travels away from the material, and the remainder goes back into the material to fuel the endothermic processes in the degradation and heating zones. The combustion products zone exists directly above the flame zone, and in this region the compounds produced during the preceding processes begin to cool and become the smoke and toxic gases associated with combustion.

This model is a very simplified view of combustion. It will be helpful, however, in the following explanation of the mechanisms by which fire retardants act.


## FIRE-RETARDANT MECHANISMS

It is possible to characterize most commercial fire retardants into one or more of four types, depending on the mechanism(s) by which they retard combustion. These four types are:

1. Intumescents
2. Heat sinks
3. Pyrolysis alterants
4. Flame poisons


### Intumescents

An *intumescent* produces or enhances the carbon structure (char) that exists on the surface of a burning material and insulates the heating and degradation zones from the heat generated in the flame zone. Anyone who

---

*One frequently hears about the tars produced in cigarettes and other smoking materials. These are liquids of relatively low molecular weight derived primarily from the polymeric tobacco and the paper sheath.

played as a child with the "black snakes" that are common during Fourth of July celebrations has seen intumescence. A more useful form, however, occurs in wood. When wood burns, its naturally porous structure carbonizes and forms a foam. This insulates the interior of the wood from the fire and is one reason why a large chunk of wood in a stove or a fireplace burns for a relatively long time. Intumescence is also significant in wood from an engineering standpoint. In a structural member, intumescence will keep a large portion of a wooden beam intact for a relatively long time during a fire. An exposed steel beam, on the other hand, will soften and lose its strength in a relatively short time, possibly causing the rapid collapse of the structure (see Figure 10.3).

## Heat Sinks

These materials act in the heating zone and/or the degradation zone by absorbing energy that would otherwise heat or decompose the polymer molecules. Water is a good example of a heat sink, as anyone who has tried to burn wet wood knows. The heat of vaporization of water is very high, and this represents energy that must be absorbed before the temperature of the substrate can rise appreciably above the boiling point of water.

It is not feasible to add water to every polymer that needs to be made less flammable. An alternative is to add a solid filler that also has an appreciable endotherm at the appropriate temperature. The material generally utilized for this purpose is alumina trihydrate,

$$Al_2O_3 \cdot 3H_2O$$

which is an intermediate product in the production of aluminum. In the 200–500°C range this material dehydrates as follows:

$$Al_2O_3 \cdot 3H_2O \rightarrow Al_2O_3 + 3H_2O\uparrow$$

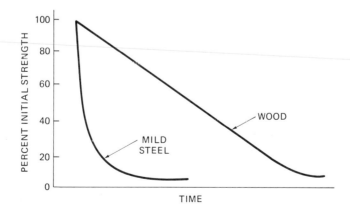

**Figure 10.3** Retention of Strength in Wood and Mild Steel during a Fire. (*Adapted from* [7].)

The reaction is strongly endothermic, with a heat of reaction equal to 197 MJ/kg. Its effectiveness and its relatively low cost make alumina trihydrate one of the highest-volume fire retardants in use today. Its traditional applications have been in reinforced polyester products, but it is gradually finding application in other areas, including thermoplastics. Its effectiveness with epoxies is illustrated in Figure 10.4.

### Pyrolysis Alterants

These are a diverse group of compounds that influence the direction of the degradation process. Recall from the six-zone model description that the flame zone is dependent on the volatile gases produced in the pyrolysis zone, since these ultimately produce a great deal of heat through oxidation in the flame zone. A pyrolysis alterant reduces the amount of volatile gases produced, by increasing the amount of char and tars. Typical pyrolysis alterants include inorganic phosphates (such as mono and diammonium phosphate), organic phosphates (such as tricresyl phosphate), and chlorinated compounds (such as chlorinated waxes). These materials are very effective. However, a drawback can be increased production of smoke as a result of the higher levels of particulates and tars.

### Flame Poisons

As their name suggests, flame poisons are effective in the flame zone. Typical flame poisons include brominated compounds and the combination of antimony trioxide and a chlorine-containing compound. These materials are free

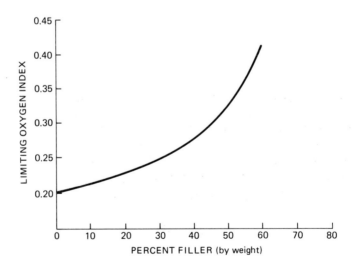

**Figure 10.4** Influence of Alumina Trihydrate on the Limiting Oxygen Index of an Epoxy. (*Adapted from* [8].)

radical scavengers that trap the free radicals ($H\cdot$, $OH\cdot$, etc.) responsible for the flame. They do this by satisfying the unpaired electron on the free radical at the expense of a molecule that is not as reactive.

The flame poison produced by combination of antimony trioxide and a chlorine-containing compound is an example of synergism. Synergism is very important in fire retardancy. It occurs when the effect of two materials together is greater than the sum of their individual effects. In this example, a small amount of antimony oxide, which by itself is not very effective in most polymers, is combined with a compound containing chlorine, which is an effective pyrolysis alterant. Their combined effect is greater than their individual effects would predict. The reason for this is not clear, although it has been speculated that an intermediate compound (antimony oxychloride) is involved.

## PROBLEMS ASSOCIATED WITH FIRE RETARDANTS

Although the purpose of fire retardants is to reduce the fire hazard presented by a material or product, they often cause problems in other areas which both manufacturers and consumers should recognize. The combustibility of a material is a fundamental characteristic, and anything used in sufficient quantity to change that characteristic is likely to change other properties as well. Problems generally occur in processing, in raw material costs, and in the retention of desired physical properties in the final product.

Processing problems can be particularly discouraging. Parameters such as melt viscosity, thermal conductivity, thermal expansion, and degradation temperature are sensitive to fillers and additives. Large adjustments (and perhaps different equipment) may be required in order to obtain reasonable production rates. For example, adding alumina trihydrate to a previously unfilled polyester resin can change the resin from a liquid of moderate viscosity into a doughy, pastelike material that may require an entirely different process to handle. On the other hand, the problem is minimal if the alumina trihydrate can be substituted directly for another filler that is already present (ground limestone, for example). Large and small processing adjustments are often a necessary part of producing a fire-retardant product.

Costs of materials are always important, and these can also change with the use of flame retardants. For example, consider the following formulation for a sheet molding compound:

| | |
|---|---|
| Unsaturated polyester resin | 31.5 parts (by weight) |
| Styrene | 3.5 parts |
| Catalyst | 0.7 parts |
| Internal release agent | 1.4 parts |
| Particulate filler (limestone) | 31.5 parts |
| Thickening agent | 1.4 parts |
| Glass fiber | 30.0 parts |

**TABLE 10.3  Costs of Various Fire Retardants (1981)**

| Material | Cost/Pound |
|---|---|
| *Additive-Type* (these are mixed with the polymer) | |
| Alumina trihydrate | $0.10 |
| Antimony oxide | $1.60 |
| Zinc borate | $0.55 |
| Tricresyl phosphate | $0.70 |
| *Reactive-Type* (these are involved in the polymerization process) | |
| Tetrabromophthalic anhydride* | $1.25 |
| Phthalic anhydride* (for comparison) | $0.42 |
| Tetrabromobisphenol A** | $1.13 |
| Bisphenol A** (for comparison) | $0.61 |

*For producing polyester resins.

**For producing epoxy resins.

One method of improving the flamespread of this formulation would be to replace the calcium carbonate with alumina trihydrate. This would increase the filler costs by a factor of 3 to 10, depending on the grades of the alumina trihydrate and the calcium carbonate. Further improvements in flamespread could be achieved by using a brominated polyester resin, but again at higher cost of materials. While the general trend in fire retardancy is toward increased costs, it is conceivable that a reduction in cost could result from a flame-retardant formulation. This could occur whenever the fire retardant replaced something that cost more, as when alumina trihydrate is used in a previously unfilled polyester. However, one would have to question in this case why a relatively inexpensive filler was not added previously, if a filled system is now acceptable. Most fire retardants are relatively expensive (Table 10.3).

Finally, the physical properties of a product may also change when a fire-retardant system is formed. The use of fillers and additives often result in composite materials with characteristics that are different from the original formulation. Properties such as modulus, strength, conductivity, hardness, thermal expansion, and hygroscopicity can be affected. Much of the problem in the development of flame-retardant formulations and products lies in the maintenance of these and other properties.

## PROBLEM SET

1. What are seven parameters that can be measured as a means of evaluating a material or product as a fire hazard? Which of these would be especially important in regard to:
   (a) Children's sleepwear?
   (b) Wall paneling?
   (c) Roofing material?
   (d) Electrical appliances?

(e) Upholstery or bedding material?

(f) Home insulation for walls, floors, and ceilings?

(g) Insulation for wiring?

2. (a) What is the limiting oxygen index?

(b) If you were given the following list of polymers with their corresponding limiting oxygen indices:

| Polypropylene | 0.175 | Epoxy | 0.198 | Nylon 6 | 0.25 |
|---|---|---|---|---|---|
| Polystyrene | 0.183 | Wool | 0.238 | Nylon 6/6 | 0.26 |

which of these would you expect to support a well-defined flame in air and which would you expect to not support one?

3. What are the six zones encountered in a burning material?

4. In which one of the zones in Problem 3

(a) Is most of the heat produced?

(b) Does the polymer degrade to lower molecular weight compounds?

(c) Do intumescent coatings work?

(d) Do heat sinks work?

(e) Do flame poisons work?

5. Define or identify the following:

(a) Intumescent coatings

(b) Alumina trihydrate

(c) Synergism

(d) Free radical

6. (a) Why does unplasticized PVC have good burning resistance?

(b) What happens to this resistance when plasticizers are added?

(c) Why do manufacturers add plasticizers to PVC?

7. What are the four basic types of fire retardants?

8. The cost figures for various fire retardants (Table 10.3) were obtained in mid-1981. Using the *Chemical Marketing Reporter*, obtain current prices for as many of these compounds as possible.

9. One method of making polymers flame-retardant is to build flame-retardant characteristics directly into the molecule. In epoxies, for example, the repeating unit may contain

as opposed to

What advantage does this have over using a brominated additive?

10. There is a great deal of controversy concerning the relevance of many flammability tests. Emmons [9] reported on a project involving a series of 27 wall-covering products, which were sent to laboratories in six countries for evaluation. Each laboratory rank-ordered the panels from the most hazardous to the least hazardous, as indicated by their standard test for that type of product. Obtain a copy of this article and summarize the results.

11. What special flammability problems might engineering applications of polymers pose in each of the following?
    (a) Submarines and space capsules
    (b) Surface ships
    (c) Mines
    (d) Airplanes

12. How does the death rate from fire in the United States compare with that in other industrialized countries? (Suggested reference: [1].)

## REFERENCES

[1] National Fire Protection Association, *National Fire Protection Association Handbook*, 14th ed., National Fire Protection Association, Boston (1976).

[2] American Society for Testing and Materials, *ASTM Standards, Part 35*, Philadelphia (1981).

[3] American Society for Testing and Materials, *ASTM Standards, Part 18*, Philadelphia (1981).

[4] Mark, H. F., S. M. Atlas, S. W. Shalaby, and E. M. Pearce, Combustion of Polymers and its Retardation, in *Flame Retardant Polymeric Materials* (M. Lewin, S. M. Atlas, and E. M. Pearce, eds.), Plenum, New York (1975).

[5] Rogers, T. H., and R. E. Fruzzett, Flame Retardance of Rubbers, in *Flame-Retardant Polymeric Materials* (M. Lewin, S. M. Atlas, and E. M. Pearce, eds.), Plenum, New York (1975).

[6] Hoke, C. E., Compounding Flame Retardance into Plastics, *SPE Journal*, vol. 29 (May 1973), pp. 36-40.

[7] American Institute of Timber Construction, *What About Fire?* Englewood, Colorado (no date).

[8] Martin, F. J., and K. R. Price, Flammability of Epoxy Resins, *Journal of Applied Polymer Science*, vol. 12, (1968), pp. 143-158.

[9] Emmons, H. W., Fire and Fire Protection, *Scientific American*, vol. 231, no. 1 (July 1974), pp. 21-27.

Many engineering applications of polymers depend on their solution properties and their ability or inability to dissolve in specific solvents. Some of these applications include the use of polymers as containers (polyethylene gas tanks), pipe (PVC sewer pipe), corrosion-resistant coatings (epoxy paints), solvent-based coatings (cellulose nitrate lacquers), adhesives (liquid phenolic resins), and in multiviscosity lubricating oils [1]. In these and other applications, many of the characteristics of polymer molecules discussed previously (chain length, crystallinity, secondary bonding, etc.) are involved. This chapter outlines some important considerations and applications.

## DISSOLVING POLYMERS

Dissolving involves taking a molecular substance (usually a solid), called the solute, and dispersing its molecules uniformly throughout the molecules of a second molecular substance (usually a liquid), which is called the solvent. For example, table sugar, (+)-sucrose:

is readily dissolved in water:

by the water's disruption of the crystalline structure of the sugar and interaction with the individual sugar molecules. Materials of low molecular weight (such as sugar) generally dissolve according to the "like dissolves like" rule. That is, polar materials tend to dissolve other polar materials and nonpolar materials tend to dissolve other nonpolar materials. Water dissolves sugar, but not grease; the components of gasoline will dissolve grease but are a poor solvent for sugar.

Dissolving involves more than this rule, however, and this is particularly evident in polymers. For example, gasoline contains $n$-heptane:

$$H—(CH_2)_7—H$$

and will readily dissolve paraffin wax:

$$H—\left[\begin{array}{c} H \ \ H \\ | \ \ \ | \\ C—C \\ | \ \ \ | \\ H \ \ H \end{array}\right]_n—H$$

At room temperature, on the other hand, it will not dissolve polyethylene, which has the same basic repeating unit as paraffin. In fact, many gasoline tanks today are made from polyethylene.

These effects can be explained with the help of the Gibbs free energy function [1]:

$$\Delta G = \Delta H - T\Delta S$$

A solution is feasible only when the change in the Gibbs free energy is negative or zero. This change is a function of both bond interactions (the "like dissolves like" rule) and the change in entropy. In materials of low molecular weight, solution can occur whether $\Delta H$ is positive or negative. When urea dissolves in water, for example, it does so endothermically ($\Delta H > 0$) and the container cools appreciably. When lithium carbonate ($Li_2CO_3$) dissolves in water, on the other hand, it does so exothermically ($\Delta H < 0$) and the container feels warm. The reason that both conditions can occur is that the change in entropy is very large when the molecules being dissolved are relatively small:

SMALL MOLECULES IN AN ORDERED
ARRAY

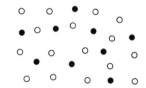

SAME MOLECULES IN A SOLUTION
(extremely disordered)
O = SOLVENT MOLECULE

Consequently $\Delta H$ can vary over a relatively wide range of values for negative values of $\Delta G$.

Because of the long-chain nature of polymers, $\Delta S$ is relatively small when they dissolve*:

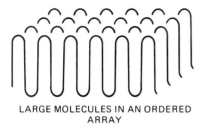

| LARGE MOLECULES IN AN ORDERED ARRAY | SAME MOLECULES IN A SOLUTION O = SOLVENT MOLECULE |

Thus the $-T\Delta S$ term will not contribute as much to $\Delta G$ as in the case of materials of low molecular weight. For solution to occur, the change in enthalpy must be negative (indicating that the solvent–solute interactions are stronger than the solvent–solvent and solute–solute bonding) or only very slightly positive (indicating the reverse). These conditions are summarized in Table 11.1.

**TABLE 11.1  Conditions for Solution to Occur**

| Material | $\Delta S$ | $\Delta H$ | Example |
|---|---|---|---|
| Low molecular weight | High | Can be greater than 0 or less than 0, but must be less than $T\Delta S$ | Urea in water ($\Delta H > 0$) |
| High molecular weight | Low | Must be less than 0 or only very slightly positive (less than $T\Delta S$) | Polystyrene in toluene |

Amorphous polymers such as polystyrene, polymethyl methacrylate, and polyvinyl chloride are soluble in many solvents even at room temperature. An interesting demonstration, for example, is to add a small amount of toluene to a polystyrene foam cup. Because of the high surface-to-volume ratio of the foam, the bottom of the cup will dissolve in a matter of seconds. A more practical application of room-temperature solubility is solvent welding, which is used extensively on some polymethyl methacrylate and polyvinyl chloride products. This involves dissolving the contact surfaces of the materials to be joined and having molecules from both surfaces interact in this solution. When the solvent evaporates, portions of these entwined molecules are trapped at the interface and bridge the two parts.

The solvent resistance of crystalline polymers (particularly at room temperature) tends to be much higher than that of amorphous polymers. This is due to the tighter structure of the crystallites, which resists penetration by solvent molecules. For example, one does not solvent-weld polypropylene pipe, because

---

*The argument in this case is the same as that presented in Chapter 8, dealing with entropy changes during melting.

the crystalline regions of isotactic polypropylene are resistant to solvents below 80° C [2]. This resistance in polypropylene and in other crystalline polymers such as high-density polyethylene and polytetrafluoroethylene is utilized in many container applications including laboratoryware, pipe, and chemical storage bottles and drums. Crystallinity is not a guarantee against room-temperature solubility, however; in some cases room-temperature solvents do exist. For example, phenol will dissolve nylon at room temperature.

## PLASTICIZERS

A middle ground exists between a solid polymer and that same polymer completely dissolved in a relatively large amount of solvent. Limited quantities of solvent can be absorbed into amorphous polymers and into the amorphous regions of crystalline polymers without creating liquids. Polymers to which solvents have been added in limited quantities are plasticized polymers. The primary effect of these solvents (plasticizers) is to lower the glass transition temperature (Chapter 8) by working themselves between the polymer chains and opening the structure somewhat. This reduces the interactions between the polymer molecules and therefore makes the material softer and more pliable.

Plasticizers are frequently added to some polymers. The commercial plasticized polymer of greatest interest today is polyvinyl chloride. In any of its applications requiring softness and flexibility, PVC is plasticized with compounds such as the esters of phthalic acid,

which are typically added in amounts ranging from 40 to 60 phr.* Much of the PVC sold today is heavily plasticized. This includes items such as shower curtains, upholstery, table cloths, and vinyl roofs for automobiles. An exception is rigid PVC used for pipe.

An important consideration in the use of plasticizers is whether they will stay mixed with the polymer or migrate to the surface. The familiar "vinyl" smell that confronts anyone who buys a new car or a shower curtain derives from the plasticizer in the polyvinyl chloride. Over time this smell diminishes, but along with it goes much of the original pliability and crack resistance of the material. Some ways to limit this migration include:

1. Choosing a plasticizer that is very compatible with the polymer
2. Choosing a plasticizer that is a large molecule
3. Choosing a plasticizer that has a relatively low volatility

*Parts by weight per hundred parts of polymer.

The esters of phthalic acid are examples of external plasticizers; that is, they are mixed with polyvinyl chloride molecules. In contrast, an internal plasticizer is one that has the same effect as an external plasticizer but is part of the polymer molecule as opposed to being an additive. Internal plasticizers are comonomers in copolymer systems, which lower the glass transition temperature by making the molecule more flexible. An example of this is the random copolymerization of butyl acrylate,

$$
\begin{array}{cc}
H & H \\
| & | \\
C & = C \\
| & | \\
H & C=O \\
 & | \\
 & O \\
 & | \\
 & C_4H_9
\end{array}
$$

with methyl methacrylate,

$$
\begin{array}{cc}
H & CH_3 \\
| & | \\
C & = C \\
| & | \\
H & C=O \\
 & | \\
 & O \\
 & | \\
 & CH_3
\end{array}
$$

A homopolymer of butyl acrylate has a $T_g$ of about $-55°C$, which implies that it is a relatively flexible molecule. Polymethyl methacrylate, of course, is a relatively rigid polymer at room temperature and its homopolymer has a much higher $T_g$ (approximately $105°C$). The copolymerization of butyl acrylate with methyl methacrylate results in a system with a lower $T_g$, given by

$$\frac{1}{T_g} = \frac{w_1}{T_{g_1}} + \frac{w_2}{T_{g_2}}$$

where $w_1$ and $w_2$ are weight fractions of the comonomers and $T_{g_1}$ and $T_{g_2}$ are the glass transition temperatures of the homopolymers. This reduction in $T_g$ causes changes in the stress–strain behavior. For example, reductions in tensile modulus and strength and an increase in elongation to break at room temperature are observed.

Similar effects are possible with an external plasticizer, but the internal plasticizer has the advantage of being fixed in the structure and not subject to migration. Other polymers that are frequently plasticized internally include polyvinyl acetate,

$$
\begin{array}{cc}
H & H \\
| & | \\
\left[ \; C \!-\! C \; \right]_n \\
| & | \\
H & O \\
 & | \\
 & C=O \\
 & | \\
 & CH_3
\end{array}
$$

and polyvinyl chloride.

## POLYMER BLENDS

In addition to mixing a polymer with a solvent of low molecular weight, it is also possible to mix (to varying degrees) two or more different types of polymer molecules or molecular segments. Methods of production in these cases range from simple mechanical blending to more complex copolymerization reactions with concomitant phase separation. These products are loosely termed polymer blends. Complete solubility in polymer blends is rare, and some degree of phase separation usually exists. This partial solubility is a major reason for their useful and interesting properties. For example, two $T_g$'s (each corresponding to a different polymer phase) are frequently found in blends of two relatively incompatible polymers.

Many important commercial polymer blends exist, including rubber-modified polystyrenes and ABS. It is interesting to note also that wood, while a natural polymer, is a rather complex polymer blend consisting primarily of crystalline cellulose and amorphous hemicellulose and lignin. For more information concerning polymer blends, see Manson and Sperling [3].

## PROBLEM SET

1. Benzene is a good solvent for natural rubber that is not crosslinked. What would happen if you put a piece of crosslinked natural rubber into a container of benzene?

2. What is a major factor that distinguishes the solution properties of polymers from those of other materials?

3. How do crystalline and amorphous polymers compare with respect to room-temperature solubility? Why?

4. How would you expect the solubility of a crystalline polymer to differ above and below its crystalline melting point? Why?

5. How would you explain the effect of crystallinity on solubility in terms of the Gibbs free energy function?

6. Distinguish between an internal and an external plasticizer.

7. Why do external plasticizers tend to migrate? What can be done to limit this?

8. How does the price of plasticizer for polyvinyl chloride (dioctyl phthalate, for example) compare with the price of polyvinyl chloride? (Suggested reference: [4].)

9. PVC is the most commonly plasticized polymer. Rigid forms contain very little plasticizer, whereas flexible varieties contain a great deal. Using the *Modern Plastics Encyclopedia* [2], compare the tensile strength, stiffness, and elongation to break of these two forms of PVC.

10. Tetrahydrofuran readily dissolves polyvinyl chloride. Why would it be a poor choice over dioctyl phthalate as a plasticizer for polyvinyl chloride?

## REFERENCES

[1] Rosen, S. L., *Fundamental Principles of Polymeric Materials*, Wiley, New York (1981).

[2] *Modern Plastics Encyclopedia*, McGraw-Hill, New York (1980–1981).

[3] Manson, J. A. and L. H. Sperling, *Polymer Blends and Composites*, Plenum Press, New York (1976).

[4] *Chemical Marketing Reporter*, Schnell Publishing Co., New York.

# introduction to mechanical properties

## chapter 12

Of the various measurements of mechanical properties made on materials, stress–strain behavior in tension is one of the most common. For many polymers this is specified by ASTM D 638, Standard Method of Test for Tensile Properties of Plastics [1]. In this test a dumbell-shaped sample is clamped in the jaws of a testing machine, and the load that is required to elongate it at a given rate is recorded. Ideally, the elongation is measured over a portion of the narrowed section of the sample (termed the *gage length*) by means of an extensometer. When low-density polyethylene is tested by this method, results similar to that in Figure 12.1 are obtained. This load versus elongation curve can be converted to a more general nominal-stress–nominal-strain curve by dividing the ordinate values by the original cross-sectional area of the narrow region of the sample ($A_0$) and dividing the abscissa values by the original gage length ($l_0$). Since $A_0$ and $l_0$ are constants*, the stress–strain curve has the same general shape as the load–elongation curve.

A nominal stress–strain curve contains much useful information about a material. This includes:

1. *Elastic modulus*: This is the slope of the initial (approximately linear) portion of the curve. In the low-density polyethylene example it is given by

---

*As illustrated in Figure 12.1, Poisson effects and plastic flow result in changing cross-sectional area and length throughout the test. By referencing loads and elongations to the *original* unstressed values, one obtains the nominal stress–strain curve as opposed to the true stress–strain curve. The discussion here is limited to the nominal case, since this is the more useful one for most design purposes.

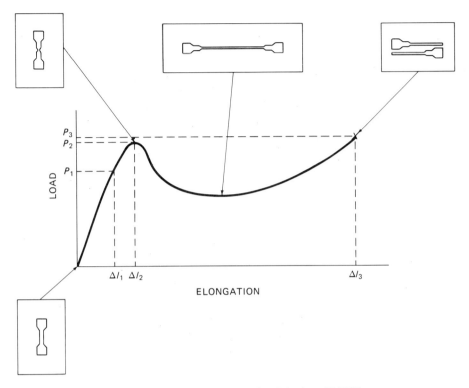

**Figure 12.1** Typical Load–Elongation Behavior of LDPE.

$$E = \frac{P_1/A_0}{\Delta l_1/l_0}$$

2. *Yield stress*: This is defined as the first stress value for which $d\sigma/d\epsilon = 0$. In the example the yield stress corresponds to $P_2/A_0$. In materials that do not usually exhibit a true yield point, an offset method is commonly used.

3. *Ultimate stress*: This is the maximum stress value. In the example the ultimate stress is $P_3/A_0$.

4. *Toughness*: This is the total area under the stress–strain curve. The units for toughness are $J/m^3$, or energy per unit volume of the original material.

When polymers are tested in tension according to ASTM D 638, four general types of behavior are frequently observed. These are:

1. Low modulus, weak, and with low toughness
2. High modulus, strong, and with low toughness
3. High modulus, strong, and tough
4. Low modulus, strong, and tough

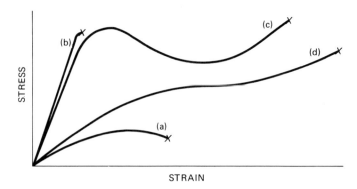

**Figure 12.2** General Types of Tensile Stress–Strain Behavior in Polymers as Determined by ASTM D 638. (a) Low-Modulus, Weak Materials with Low Toughness. (b) High-Modulus, Strong Materials with Low Toughness. (c) High-Modulus, Strong, Tough Materials. (d) Low-Modulus, Strong, Tough Materials.

The characteristics of each type are summarized in Figure 12.2.

Many of the characteristics of polymers that were discussed in earlier chapters are influential in determining tensile behavior of a given polymer. By way of review, these characteristics include:

1. The structure of the repeating unit (as it influences the secondary bonding)
2. The flexibility of the molecules
3. The molecular weight
4. The presence or absence of crosslinks
5. Crystallinity and the orientation of molecules
6. Temperature and its relationship to $T_g$ and $T_m$

It is specific combinations of the above that contribute to the responses in Figure 12.2.

For example, one polymer that has a low modulus, is relatively weak, and has low toughness is atactic polypropylene:

$$\left[\begin{matrix} H & CH_3 \\ | & | \\ C - C \\ | & | \\ H & H \end{matrix}\right]_n$$

From the discussions in previous chapters one should recall the following about this material:

1. The bonding between the molecules is van der Waals (the weakest of the three possibilities).

2. The flexibility of the chain is fairly high (it is lower than that of polyethylene because of the added steric hindrance that the methyl group presents, but much higher than that of polystyrene, for example). The $T_g$ is about $-18°$C.

3. The atactic nature of the chain makes it difficult for the material to crystallize.

Collectively (1), (2), and (3) mean that atactic polypropylene is composed of rather weakly bonded, relatively flexible chains tied together with few crystals. When molecules of this nature are stressed, they provide relatively little resistance; consequently atactic polypropylene is flexible and weak at room temperature. It has limited practical use. Isotactic polypropylene, on the other hand, is a widely used, stiff, strong, and tough polymer. This difference derives largely from the improved crystallinity.

Similar analyses can be made for other polymers such as polystyrene, polyvinyl chloride, epoxies, and natural rubber. The reader is encouraged to pursue these analyses, particularly in the course of studying the structures of the polymers. The following discussion of the general effects of molecular parameters on modulus and strength is intended as an aid to this.

## MODULUS

The modulus of a polymer is related to both its intramolecular and its intermolecular bonding. Given molecules in an unstressed state,

an applied stress will move them in relation to one another and disrupt some of the intermolecular bonding:

If this intermolecular bonding is stiff and strong, it will provide significant resistance to the stress. As the chains move, some straining and rotation of the intramolecular bonds in the molecule will occur, and this also resists the stress. Thus those polymers possessing the strongest intermolecular bonds and the stiffest chains tend to have the highest moduli.

Any molecular feature that increases the strength and stiffness of the secondary bonds and/or reduces the flexibility of the chains will tend to increase modulus. Crystals increase modulus through both of these mechanisms, since the crystalline lattice involves molecules that are relatively close to one another. The bonds are stronger and stiffer, since the intermolecular distances are shorter, and the flexibility of the chains is hindered by adjacent molecules. The orientation of the crystals also influences modulus. In an oriented polymer the modulus tends to be highest in the orientation direction because a greater proportion of the load is carried by stiff, strong covalent bonds in the backbone of the molecules. In other directions greater proportions of the load are carried by the weaker, more compliant secondary bonds, and the moduli in these directions are therefore lower.

In the same general fashion, crosslinks also increase modulus, since they supplement the secondary bonding and reduce molecular mobility. However, the degree of increase depends on the flexibility of the molecules before crosslinking. For example, crosslinking a material whose $T_g$ is already well above the use temperature will have less effect on modulus than for one whose $T_g$ is below the use temperature. In other words, if the molecules are already very stiff, the crosslinks will have less additional effect.

Stiff or flexible backbone groups (such as a benzene ring or an oxygen linkage) can affect the mobility of the molecules and therefore the modulus of the material. The arguments in this case are the same as those for the influence of backbone groups on $T_g$. Similar effects are also noted for large or bulky side groups that hinder rotation about carbon–carbon single bonds.

Modulus is also somewhat dependent on molecular weight. In polymers this dependency can be visualized to a large extent as an increasing degree of molecular entanglement. As chain length increases, the molecules become more entwined; and these entanglements act as temporary crosslinks when they are stressed. The effects are greatest for short lengths, and changes in length are less influential when the chains are already long and entangled. Thus a modulus versus molecular weight curve tends to have the shape discussed in Chapter 3.

Temperature can also have a strong influence on the modulus of a polymer, particularly in and near the transition regions. The modulus of an amorphous polymer below its $T_g$ will decrease slightly with increasing temperature.\* As $T_g$ is approached, however, the modulus will decline at an extremely rapid rate, as illustrated in Figure 12.3. Above $T_g$ the polymer will be either a fluid or a rubbery material, depending on the presence or absence of crosslinks. If no crosslinks are present, a fluid results whose viscosity decreases with increasing temperature. If crosslinks are present, the polymer becomes a rubbery material rather than a fluid, since the crosslinks prevent complete slippage of molecules past one another. The characteristics of rubbery polymers are discussed in Chapter 14.

---

\*Secondary transitions can also occur below $T_g$, resulting in substantial changes in modulus within a narrow range of temperatures. These effects are ignored in the present discussion.

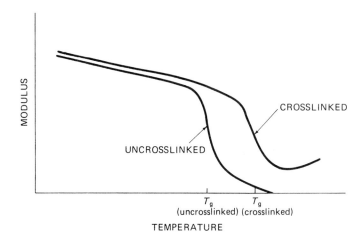

**Figure 12.3** Modulus versus Temperature for Amorphous Polymers.

In semicrystalline polymers substantial changes in modulus can occur at both $T_g$ and $T_m$. The relative magnitudes of these changes depend on the relative amounts of amorphous and crystalline material, since both regions contribute to the modulus of the bulk polymer. Below $T_g$ in a semicrystalline polymer both the crystalline and amorphous regions are rigid, and the resulting modulus is relatively high. At $T_g$ the crystals remain rigid, but the nature of amorphous regions changes from glassy to rubbery. This results in a lower modulus for the bulk polymer, but one that is still relatively high because of the crystals. At $T_m$ the crystals melt, and the polymer (if uncrosslinked) becomes a fluid. These effects are illustrated in Figure 12.4.

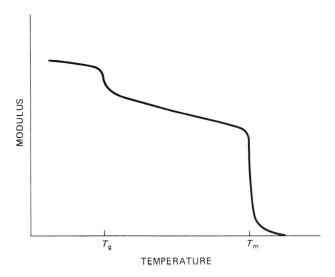

**Figure 12.4** Modulus versus Temperature for a Semicrystalline Polymer.

## STRENGTH

Predicting strength is difficult in most materials, and polymers are no exception. Although the theory is not well developed, it is known that defects and the resulting stress concentrations are important and that both the primary and the secondary bonds can influence the strength of polymers [2]. The mechanical equivalent strength of any bond can be roughly estimated by

$$E = Fd$$

where:

$E$ = the bond dissociation energy (J/bond)

$F$ = the load required to break the bond (N)

$d$ = 0.15 nm for covalent bonds

   0.3 nm for hydrogen bonds

   0.3 nm for permanent dipole bonds

   0.4 nm for van der Waals forces

For example, a carbon–carbon single bond with a dissociation energy of $5.8 \times 10^{-19}$ J requires approximately

$$F = \frac{5.8 \times 10^{-19} \text{ J}}{1.5 \times 10^{-10} \text{ m}}$$

$$= 39 \times 10^{-10} \text{ N}$$

to break. In a similar fashion it can be seen that the mechanical strengths of secondary bonds are much lower than those of covalent bonds: approximately $1 \times 10^{-10}$ N for hydrogen bonds, and $0.2 \times 10^{-10}$ N for van der Waals forces [2].

These calculations suggest that if all the molecules in a polymeric material could be aligned, if the stress was in the direction of the alignment of the primary bonds, and if all the molecules were loaded equally, then a very strong material would result. For example, consider polyethylene molecules in a crystalline lattice:

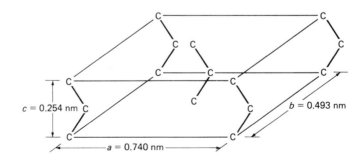

If the molecules did not pull out of this lattice, the tensile strength of this structure in the $c$ direction would be approximately

$$\text{Strength} = \frac{\text{Load}}{\text{Area}} = \frac{\left(\begin{array}{c}\text{Number of}\\\text{covalent bonds}\end{array}\right)\left(\begin{array}{c}\text{Strength of a}\\\text{covalent bond}\end{array}\right)}{\text{Area}}$$

$$= \frac{[(\tfrac{1}{4} \cdot 4 + 1) \text{ bonds}] (39 \times 10^{-10} \text{ Newtons/bond})}{(7.40 \times 10^{-10} \text{ m}) (4.93 \times 10^{-10} \text{ m})}$$

$$= 2.14 \times 10^{10} \text{ N/m}^2$$

$$= 21,400 \text{ MPa}$$

This value is extremely high. When commercial polyethylenes are tested in accordance with ASTM D 638 [1], ultimate strength values are typically on the order of 4.2 to 15.9 MPa for low-density polyethylene and 21.4 to 37.9 MPa for high-density polyethylene [3]. The highest values recorded for oriented polyethylene filaments are only approximately 1379 to 1655 MPa [2]. The difference between these values and that for the theoretical, unidirectional case occurs because no polymer can have all its chains arranged as described above. Chain ends, branches, amorphous regions, crystal size, and other crystal orientations contribute to this difference between theoretical and actual values.

Some typical values of tensile modulus and strength for various bulk polymers are listed in Table 12.1. All these values were obtained in accordance with ASTM D 638 [1]. It is interesting to note that the tensile moduli and strengths of most polymers are below approximately 5000 MPa and 100 MPa, respectively. In comparison, the tensile moduli and ultimate strengths of low-carbon steels are approximately 200,000 MPa and 500 MPa, respectively.

**TABLE 12.1  Tensile Modulus and Strength of Various Synthetic Polymers**

| Polymer | Tensile Modulus (MPa) | Ultimate Tensile Strength (MPa) |
|---|---|---|
| Polymethyl methacrylate | 2400–3100 | 55–75 |
| Polytetrafluoroethylene | 400–550 | 15–35 |
| Nylon 6 (50% rh) | 700 | 70 |
| Polycarbonate | 2400 | 65 |
| Polyethylene terephthalate | 2750–4150 | 60–70 |
| Low-density polyethylene | 100–250 | 5–15 |
| High-density polyethylene | 400–1250 | 20–40 |
| Polypropylene | 1150–1550 | 30–40 |
| Polystyrene | 2400–3350 | 35–55 |

(*Adapted from* [3].)

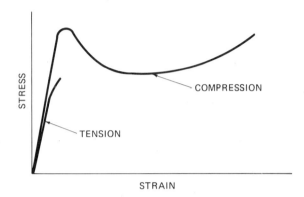

**Figure 12.5** Stress–Strain Behavior of Polystyrene in Tension and Compression. (*Adapted from* [4], p. 261, by courtesy of Marcel Dekker, Inc.)

## STRESS-STRAIN BEHAVIOR IN OTHER TESTING MODES

Although stress–strain in tension is a useful test for comparing materials, it is only one of many mechanical tests that are used. The behavior of a polymer is a function of the testing mode; that is, modulus, strength, and toughness may differ considerably for a given polymer in tension as opposed to compression, bending, or shear. For example, polystyrene is a brittle material in tension, but in compression it can be rather ductile. This difference in behavior is illustrated in Figure 12.5.

As is the case for stress–strain in tension, ASTM tests exist for determining the compressive, bending, and shear properties of polymers. Most of these tests are included in Parts 35 and 36 of the *ASTM Standards.* Table 12.2 lists several of the more common tests. In addition to these, others are frequently utilized to characterize the mechanical behavior of polymers for specific purposes. Some of

**TABLE 12.2  Common ASTM Tests Used to Determine the Tensile, Compressive, Shear, and Bending Properties of Polymers (in ASTM Standards, Parts 35 and 36)**

| Testing Mode | ASTM Designation | Test |
|---|---|---|
| Tension | D 638 | Tensile Properties of Plastics |
| | D 1623 | Tensile Properties of Rigid Cellular Plastics |
| | D 882 | Tensile Properties of Thin Plastics Sheeting |
| Compression | D 695 | Compressive Properties of Rigid Plastics |
| | D 1621 | Compressive Strength of Rigid Cellular Plastics |
| Shear | D 732 | Shear Strength of Plastics |
| Bending | D 790 | Flexural Properties of Plastics |

these include hardness, heat distortion temperature, fatigue, tear resistance, impact strength, abrasion resistance, creep, and stress relaxation. Several of these properties are discussed in the sections that follow.

## HEAT DISTORTION TEMPERATURE

This test is specified by ASTM D 648 [1]. It utilizes a bar (1.27 cm $\times$ 1.27 cm $\times$ 12.7 cm) in a three-point loading configuration as shown:

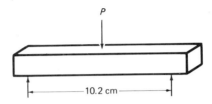

The load, $P$, is such that either 1.82 MPa or 0.46 MPa is the maximum stress in the bar.* While loaded, the bar is heated uniformly a rate of 2°C/min, and the heat distortion temperature is defined as that temperature at which the midspan deflection is 250 $\mu$m.

The practical implications of this test are obvious, since it supplies a rough estimate of the upper use temperature for materials as far as load-bearing capacity is concerned. For glassy polymers the heat distortion temperature is related to the glass transition temperature, whereas for crystalline polymers it may be related to both $T_g$ and $T_m$. Other modes such as tension and shear are also used.

## FATIGUE

When polymers are subjected to oscillating or varying loads, they fail at much lower stresses than their static test results indicate. This phenomenon is fatigue, and fatigue life is defined as the number of cycles (at a specified maximum stress) which a material can withstand before breaking. Fatigue life is a function of stress level, so data are commonly presented as in Figure 12.6. The stress level, $\sigma_1$, is the endurance limit and represents the stress below which the material can withstand an unlimited number of cycles. As noted by Nielsen [4], this value is often between 20% and 40% of the static tensile strength. In any design involving vibratory stresses, the stress levels should obviously be kept below these values.

---

*1.82 MPa is usually used for very rigid polymers, whereas 0.46 MPa is usually used for the more flexible ones.

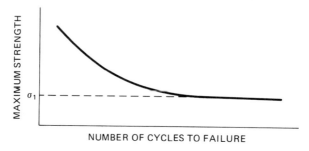

**Figure 12.6** Fatigue Strength of a Typical Polymer.

Fatigue is measured by a number of different tests, and results from different tests are not always comparable. The mechanisms of fatigue and the relationships between molecular parameters and endurance limits are not well known at this time. Although there are some good references dealing with fatigue, relatively little is available dealing with polymers. See Nielsen [4] or Hertzberg and Manson [5] for additional information.

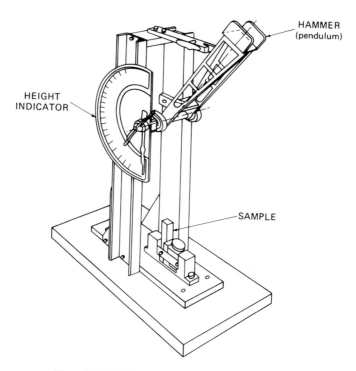

**Figure 12.7** IZOD Impact Tester. (*Adapted from* [1].)

**TABLE 12.3 Impact Strength of Various Polymers (IZOD—23°C)**

| Polymer | Impact Strength (J/m of notch) |
|---|---|
| Polymethyl methacrylate | 16–21 |
| Polytetrafluoroethylene | 160 |
| Nylon 6 (0.2% moisture content) | 43–160 |
| Polycarbonate | 854 |
| Polyethylene terephthalate | 13–35 |
| Low-density polyethylene | no break |
| High-density polyethylene | 27–1068 |
| Polypropylene | 21–53 |
| Polystyrene | 13–21 |

(*Adapted from* [3].)

## IMPACT STRENGTH

Impact strength is the energy required to break a material using very rapid loading rates. It differs from toughness largely in that the latter property does not necessarily require rapid loading. There are several tests used to measure the impact strength of polymers, but one of the most common is the IZOD test [1]. In this test a small, notched cantilever beam is broken with a pendulum hammer (Figure 12.7). The energy required to break the sample is determined from the height the pendulum reaches after breaking the sample.

Table 12.3 lists typical values for the impact strengths of some polymers as determined by the notched IZOD test. The values for a given polymer can vary over a wide range because impact strength is strongly sensitive to parameters other than those derived directly from the structure of the repeating unit. These include crystallite size, molecular weight, temperature (particularly in relation to $T_g$), molecular orientation, and the presence of plasticizers, fillers, and reinforcing agents.

## PROBLEM SET

1. Briefly describe how the following influence a polymer's modulus:
   (a) Crystals
   (b) Crosslinks
   (c) Temperature
   (d) Plasticizers
   Cite an example for each case.
2. Nylons are considered stiff, strong, and tough polymers at room temperature. What characteristics in nylons contribute to this behavior?

3. Polystyrene is considered a rigid, brittle polymer at room temperature. What characteristics in polystyrene contribute to this behavior?

4. What is a molecular entanglement? How does it influence modulus?

5. Distinguish between toughness and impact strength.

6. Why do the moduli of most polymers decrease with increasing temperature in regions other than where $T_g$ and $T_m$ occur?

7. The following data for nylon 6 were taken from *Modern Plastics Encyclopedia* [3]:

Tensile strength: 11,800 (dry); 10,000 (50% relative humidity) (psi)

Tensile modulus: $3.8 \times 10^5$ (dry); $1.0 \times 10^5$ (50% relative humidity) (psi)

IZOD impact strength: 0.6–1.0 (dry); 3.0 (50% relative humidity) (ft-lb/in of notch)

Explain the above moisture sensitivity of the results in terms of the repeating unit of nylon 6. Would you expect the same effects with polyethylene?

8. In regard to Figure 12.3, why does the modulus of the crosslinked polymer not drop to zero above $T_g$?

9. Polytetrafluoroethylene has very stiff chains and high crystallinity, but its modulus is relatively low. Explain this apparent contradiction.

10. What is the heat distortion temperature?

11. Calculate the loads required to obtain 1.82 MPa and 0.46 MPa in the heat distortion test. Recall that stress = moment/section modulus.

12. Many standard mechanical tests exist for polymer-based materials and products. Some of them are described in the *ASTM Standards*, Parts 35 and 36. Obtain copies of these manuals and list some of the mechanical tests they contain.

## REFERENCES

[1] American Society for Testing and Materials, *ASTM Standards, Part 35*, Philadelphia (1981).

[2] Mark, H., Strength of Polymers, in *Polymer Science and Materials*, (A. V. Tobolsky and H. Mark, eds.), Wiley-Interscience, New York (1971).

[3] *Modern Plastics Encyclopedia*, McGraw-Hill, New York (1980–1981).

[4] Nielsen, L. E., *Mechanical Properties of Polymers and Composites*, vols. 1 and 2, Marcel Dekker, Inc., New York (1974).

[5] Hertzberg, R. W., and J. A. Manson, *Fatigue of Engineering Plastics*, Academic Press, New York (1980).

# viscoelasticity

## chapter 13

Traditional engineering materials are generally considered either elastic or viscous. The behavior of an elastic material is governed by Hooke's law:

$$\sigma = E\epsilon$$

where $\sigma$ is the stress, $\epsilon$ is the strain, and $E$ is the elastic modulus. Since this is an equation of the form $y = mx$, it can be represented graphically as shown in Figure 13.1. This response is independent of the rate of strain, $d\epsilon/dt$. In other words, time is not a variable. No matter how fast or how slowly an elastic material is deformed, the stress–strain curve in Figure 13.1 remains the same.

Viscous materials, on the other hand, behave according to the Newtonian relationship

$$\sigma = \eta\left(\frac{d\epsilon}{dt}\right)$$

where $\eta$ is the viscosity and $d\epsilon/dt$ is the strain rate. This is much different from elastic behavior. Instead of stress being a function of strain and independent of strain rate, it is a function of strain rate and independent of strain. A typical stress–strain curve for a viscous material is shown in Figure 13.2 Time is a variable in this case because of the strain rate term. As the strain rate is increased, the stress level required to maintain the rate of strain also increases.

There are no perfectly elastic or perfectly viscous materials. All materials are to varying degrees viscoelastic, which means that stress is a function of both strain and strain rate. Stress–strain curves in this case are more complex, as shown in Figure 13.3. In the viscoelastic case the modulus ($\sigma/\epsilon$) is a function of

**105**

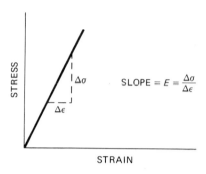

Figure 13.1 Stress–Strain Behavior of an Elastic Material.

the rate of strain; that is, the stress required for a given strain depends on the time over which the strain develops.

In design it is more convenient to work with univariate functions (the elastic and viscous cases) than multivariate ones (the viscoelastic case). Thus when one of the two variables (either strain or strain rate) contributes little to the stress, it is ignored, and the material is considered either elastic or viscous rather than viscoelastic. Under most conditions steel and many other common materials are considered elastic. The effect which time has on their stress–strain response is small enough to be ignored. On the other hand, water and oil are considered to be viscous materials, since their elastic contributions are relatively small compared to their viscous response. With synthetic polymers, unfortunately, one frequently cannot make these assumptions.

Traditionally, viscoelastic effects in solid polymers are observed in four situations:

1. Creep, in which a polymer is stressed at a constant level for a given time, and the strain increases during that time period.
2. Stress relaxation, in which a polymer is strained at a constant level for a given time, and the stress relaxes or diminishes during that time period.

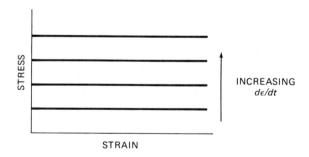

Figure 13.2 Stress–Strain Behavior of a Viscous Material.

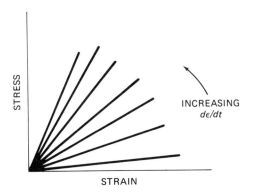

**Figure 13.3** Stress–Strain Behavior of a
Viscoelastic Material.

3. Stress–strain behavior at strain rates of approximately 0.01 to 100 m/m/ minute, in which the modulus increases with increasing strain rate.*
4. Vibrational loading, in which a material is subjected to an oscillating strain, and the resulting stress is out of phase with the strain.

Collectively, these effects cover the entire time range of practical interest for most polymer applications. Creep, on the one extreme, involves strains that develop over a period of hours or years, while vibrational effects, on the other, can involve strains that develop and are reversed in microseconds.

For many of the applications of polymers, long-term creep and stress relaxation are very important. The remainder of this chapter is limited to a discussion of these two effects. Readers interested in the shorter term effects in polymers should examine references [1]–[3], noted at the end of the chapter.

## CREEP

In a creep test a material is subjected to a constant nominal stress (usually tensile), and the change in length ($\Delta l$) that occurs over time is monitored. This change in length is divided by the original gage length ($l_0$) to yield a time-dependent strain, $\epsilon(t)$. These can be plotted as a function of time as shown in Figure 13.4. The creep modulus, $E(t)$, is defined as

$$E(t) = \frac{\sigma_0}{\epsilon(t)}$$

and is, of course, a function of time. The response in Figure 13.4 is characteristic

*These effects are usually observed in a testing machine in a laboratory, although loading rates of this magnitude are relevant to many applications. For example, an orthopedic prosthesis (such as a hip joint) is subjected to strain rates in this range.

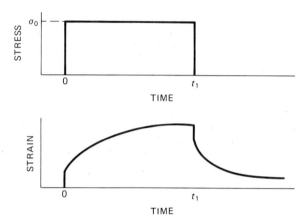

**Figure 13.4** Creep Response of a Viscoelastic Material.

of viscoelastic materials; that is, there is an initial elastic response, followed by a time-dependent strain. Upon release of the stress, the elastic portion of the strain is recovered immediately, followed by a delayed recovery of a portion of the additional strain.

In comparison, an elastic material has a creep response that is independent of time, as shown in Figure 13.5. No additional strain occurs after the initial loading. When the stress is released, all of the strain is immediately recovered. A viscous material, on the other hand, responds as shown in Figure 13.6. In contrast to the other two cases, there is no initial elastic response; when a viscous material is stressed, only time-dependent strain occurs. Upon release of the stress, there is no elastic or delayed recovery in a viscous material.

If a viscoelastic material is tested in creep for sufficient time, failure will result, as is shown in Figure 13.7. This is called *creep rupture*, and it is a useful tool for developing allowable stress levels in design. If this test is repeated at several stress levels, a plot of stress level versus time to failure can be developed

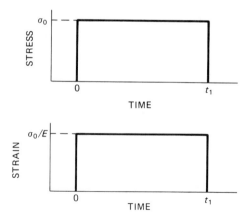

**Figure 13.5** Creep Response of an Elastic Material.

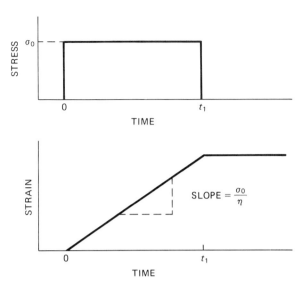

**Figure 13.6**  Creep Response of a Viscous Material.

(Figure 13.8). The stresses in Figure 13.8 can then become working stresses for design equations, after the desired design life of the part is determined and suitable safety factors are applied.

Creep rupture tests are frequently run at temperatures where thermal degradation is significant. In these cases the data can provide an accurate estimate of the elevated-temperature life of a material when it is subjected to mechanical stress. This differs from the U.L. temperature index (Chapter 9), which does not consider the effects of mechanical stress on degradation. Since these effects are significant, creep rupture data can be more relevant than the temperature index in some cases. These applications could include polymeric pipes conveying hot, pressurized liquids or polymeric automobile engine components. Examples of design problems using creep modulus and creep rupture data can be found in the *Modern Plastics Encyclopedia* [4].

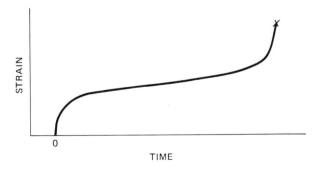

**Figure 13.7**  Creep Test to Failure in a Viscoelastic Material.

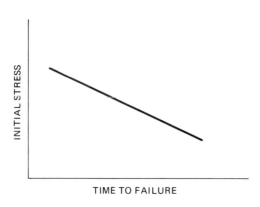

**Figure 13.8** Creep Rupture Data on a Log–Log Scale.

## STRESS RELAXATION

Stress relaxation is similar to creep in that both are generally run on the same time scale (i.e., on the order of hours to years). Mechanically, however, stress relaxation is the opposite of creep, since the strain is held constant ($d\epsilon/dt = 0$) and the stress required to maintain the strain is monitored. A stress-relaxation test is begun by very rapidly stretching a specimen to a given elongation. The load required to maintain that elongation is then measured over time. These parameters are converted to nominal stress and strain and plotted as a function of time, as shown in Figure 13.9. The relaxation modulus, $E(t)$, is defined as

$$E(t) = \frac{\sigma(t)}{\epsilon}$$

where $\epsilon$ is the applied strain and $\sigma(t)$ is the residual stress at time $t$. Another parameter, the relaxation time, $\lambda$, is defined as the time required for the stress to decay to $1/e$ of its original value ($\sigma_0$ in Figure 13.9).

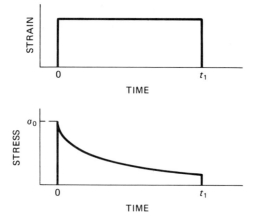

**Figure 13.9** Stress Relaxation in a Viscoelastic Material.

It is perhaps more difficult to see the importance of stress relaxation than to see the importance of creep; but there are a number of practical applications. Consider, for example, a polymeric gasket for a pressurized container. When the bolts on the container lid are torqued, the gasket is compressed or strained to a given level. Because of this applied strain, a stress develops in the material that is directly proportional to the modulus. This stress prevents the container from leaking. Since the modulus of the gasket material is a function of time, the stress level in the gasket will diminish over time. This is particularly true at high temperatures. The container will eventually leak because of this, and at that point the gasket must be retightened or replaced. There are many other practical examples that involve a constant strain and in which performance depends on a certain stress level in the material. Vacuum hoses that loosen and strings on a musical instrument whose tension decreases with time are other common examples.

## PARAMETERS AFFECTING CREEP AND STRESS RELAXATION IN POLYMERS

Both creep and stress relaxation involve the time-dependent motion of molecules while they are stressed. This motion can assume several forms [3], depending on the time involved. For very short loading times, only a distortion of bond distances occurs. This yields the lowest strains and therefore the highest creep and relaxation moduli for a given system. As the loading time increases, however, larger-scale molecular uncoiling motion becomes possible. Initially, these movements involve small segments of the molecule; but at longer loading times, movement and relocation of much larger segments become possible. In addition, stress-induced recrystallization can occur in semicrystalline polymers. All these effects involve the limited slippage of molecular segments, which tends to increase the strain (or relax the stress) observed in the bulk polymer and, therefore, lowers the creep and relaxation moduli. The largest-scale molecular motion possible in polymers is the complete slippage of molecules past one another. In solid polymers this requires the longest loading times to occur and leads ultimately to the failure of the material [3].

The degree to which these effects (and therefore creep and stress relaxation) occur in a given system depends on many parameters other than loading time and the initial stress level. These parameters include

1. Crystallinity
2. Crosslinking
3. Molecular weight
4. Temperature

The influence of each of these on creep and stress relaxation is briefly discussed next.

## Crystallinity

The effects of crystallinity on creep can be rather complex because of slip planes and recrystallization effects that can occur under stress. As a first approximation, however, crystals tend to reduce creep (particularly in polymers of low crystallinity), since the molecules in crystals have shorter bond distances (and, therefore, stronger intermolecular bonds) than those in the amorphous regions. Similarly, crystal orientation is also influential, with the lowest creep usually occurring in the orientation direction.

## Crosslinking

Crosslinking reduces creep, since it limits the amount of slippage that can occur between two molecules. When they are stressed, crosslinked molecules will tend to uncoil and slide past one another somewhat in regions between the crosslinks. Once the "slack" is removed and the crosslinks are stressed, however, no additional slippage is possible (unless covalent bonds are broken). How quickly this tightening occurs depends on the number of crosslinks and their positions in the polymer network. Figure 13.10 illustrates the general effect of increased crosslinking on creep.

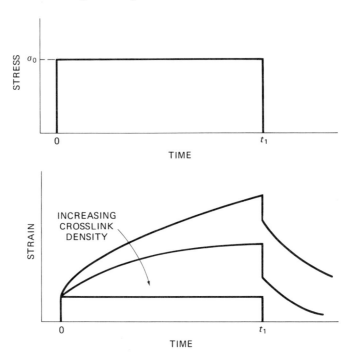

**Figure 13.10** The Influence of Crosslink Density on Creep (Amorphous Polymer). (*Adapted from* [1].)

## Molecular Weight

Increasing the molecular weight of a given polymer also tends to reduce creep. This is due to the higher degree of entanglement that accompanies an increase in chain length. The entanglements act as temporary (labile) crosslinks, restricting the slippage that can occur in a given amount of time. The general trend of creep response versus molecular weight is summarized in Figure 13.11.

## Temperature

An increase in temperature usually weakens the intermolecular bonding and will, therefore, increase creep. This effect is often used to predict creep and stress relaxation at very long times that would be impractical to measure otherwise. To find the creep modulus of a material for a given stress at, for example, 5 years and 23° C, a test is run for a much shorter time at an elevated temperature. This useful procedure is called time-temperature superposition, and the details of this are available in references [2] and [3].

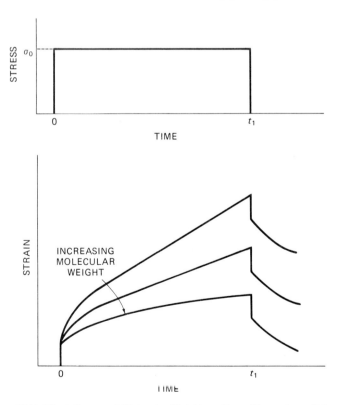

**Figure 13.11** The Influence of Molecular Weight on Creep (Amorphous Polymer). (*Adapted from* [1].)

The positions of $T_g$ and $T_m$ in relation to the use temperature of a polymer can be very important in regard to its creep response. Many amorphous polymers, for example, will exhibit very high creep rates at temperatures near $T_g$. Alternatively, a crystalline polymer near its $T_m$ can experience stress-induced recrystallization and exhibit a very high creep rate compared to a polymer whose $T_m$ is far removed from its use temperature. The sensitivity of creep to parameters such as molecular weight and crosslinking also depends to a large extent on the relationship between the application temperature and $T_g$ or $T_m$. For example, increases in molecular weight and degree of crosslinking do not usually have large influences on creep behavior at temperatures well below $T_g$. On the other hand, they do exert strong influences at and above this temperature.

## PROBLEM SET

1. What is creep? How do polymers and metals differ with respect to this property?
2. How do the following polymer characteristics affect creep?
   (a) Crosslinks
   (b) Crystals
   (c) Molecular weight
3. Draw general creep curves for:
   (a) An ideal elastic material
   (b) An ideal viscous material
   (c) A viscoelastic material
   Be sure to label the axes.
4. The following creep data for a nylon 6 product were taken from the *Modern Plastics Encyclopedia* [4]:

| Initial Applied Stress (psi) | Creep Modulus $\times$ $10^{-3}$ (psi) | | | |
|---|---|---|---|---|
| | 1 hr | 10 hr | 100 hr | 1000 hr |
| 500 | 106 | 97 | 89 | 77 |
| 1000 | 95 | 83 | 72 | 63 |
| 1500 | 78 | 68 | 59 | 51 |

How do these stress levels compare with the ultimate tensile strength of nylon 6? On this basis can you conclude that creep is important only at high stress levels?

5. What is stress relaxation? How does it differ from creep?
6. Draw general stress-relaxation curves for:
   (a) An ideal elastic material
   (b) An ideal viscous material
   (c) A viscoelastic material
7. Of what practical importance is stress relaxation?
8. The following flexural creep data were obtained for nylon 6 reinforced with 30–35% glass fiber [4]:

| Initial Applied Stress (psi) | Creep Modulus × 10⁻³ psi | | | | | |
|---|---|---|---|---|---|---|
|  | 1 hr | 10 hr | 30 hr | 100 hr | 300 hr | 1000 hr |
| 5000 | — | 660 | 570 | 550 | 540 | 520 |

The same material has a modulus of $1 \times 10^6$ psi in a static flexural test. Calculate the additional deflection that occurs as a result of creep during the first 1000 hours of loading of a simply supported beam made of the above material and carrying a constant load of 100 lbs/ft as shown:

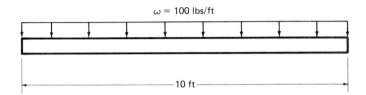

Express this additional deflection as a percentage of the initial deflection in the static case. Recall that the maximum deflection for a uniformly loaded beam is given by

$$y_{max} = \frac{5\omega l^4}{384EI}$$

where:

$y_{max}$ = maximum deflection
$\omega$ = load
$l$ = span
$E$ = modulus
$I$ = moment of inertia

**Answer:**
92%

9. According to the *Modern Plastics Encyclopedia* [4], it is safe to extrapolate creep data up to 1 decade of time. That is, data up to 100 hours may be extrapolated to 1000 hours. Using this assumption, calculate the additional deflection in the beam in Problem 8 for 10,000 hours.

# REFERENCES

[1] Rosen, S. L., *Fundamental Principles of Polymeric Materials*, Wiley, New York (1981).

[2] Nielsen, L. E., *Mechanical Properties of Polymers and Composites*, vols. 1 and 2, Dekker, New York (1974).

[3] Alkonis, J. J., W. J. MacKnight, and M. Shen, *Introduction to Polymer Viscoelasticity*, Wiley, New York (1972).

[4] *Modern Plastics Encyclopedia*, McGraw-Hill, New York (1978–1979).

# rubbery behavior

## chapter 14

Polymers that are rubbery at their use temperature are often called elastomers. These are a very special group of polymers with specific molecular characteristics that result in their unique mechanical behavior. For this reason they are treated separately in this text.

To be considered (ideally) elastomeric, a polymer should possess the following mechanical behavior [1]:

1. It must be rapidly stretchable to very high extensions (on the order of 500%).
2. It must possess high tensile strength when fully stretched; that is, the stress–strain curve should have shape

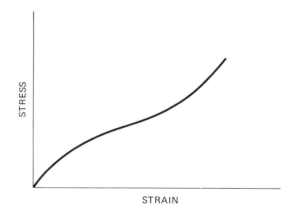

as opposed to shape

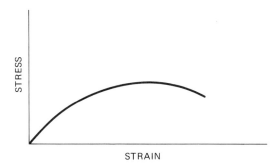

3. It must snap back when the stress is released.
4. It must retract completely with no permanent set.

The last three requirements are equivalent to stating that the strain energy is stored elastically in the material (i.e., none is converted into heat).

When this behavior is observed, certain molecular and environmental conditions usually exist [1]:

1. The material is a polymer.
2. The material is amorphous.
3. The temperature is above $T_g$.
4. The material is lightly crosslinked.

The reasons for these conditions become evident when one considers the large extension ratios that elastomers must exhibit. In the unstressed state the elastomer consists of coiled, relatively flexible molecules entangled with and lightly crosslinked to one another:

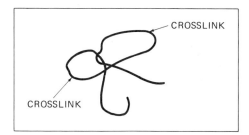

When stressed (and, in particular, when stressed to high elongations), these molecules are stretched into lower entropy conformations:

until the crosslinks prevent further significant elongation. When the stress is released, the stored strain energy restores the molecules to their original conformations:

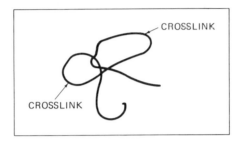

Unless all the conditions listed previously exist, the material usually does not exhibit this type of behavior. For example, short chains would have little opportunity to coil and uncoil, and a material containing them would have a low extension to break. If crystals were present (or the temperature was below $T_g$), the molecules would have restricted mobility and the bulk material would be inflexible. Finally, without crosslinks some permanent flow would occur under stress and the molecules would not return to their original state.

## PROPERTIES OF IMPORTANT COMMERCIAL ELASTOMERS

Because of the variation that exists in the structures of repeating units, many types of elastomers are commercially available. Some of these are listed in Table 14.1. As this table indicates, many of the common elastomers are copolymers, as opposed to homopolymers. There are several reasons for this. In butyl rubber, for example, a relatively small amount of isoprene comonomer is utilized. The purpose of this comonomer is to provide sites for crosslinking reactions, since the butyl repeating unit is not receptive to either sulfur or peroxide crosslinking. This is also done with acrylate elastomers (using 2-chloroethyl vinyl ether). Each of these systems involves a large proportion of one comonomer and a small proportion of another. Some copolymer elastomers contain large proportions of more than one comonomer. In ethylene-propylene rubber, for example, the propylene content can vary from approximately 30 to 60 mole percent. The purpose of this is to create an amorphous material. Polyethylene and polypropylene homopolymers are both crystalline, but the random copolymer has little tendency to crystallize when substantial quantities of each comonomer are present. The copolymer structure is irregular because the propylene units are positioned at random in the molecules. This discourages crystal formation because of the dissimilar nature of neighboring molecules.

Some of the properties of butyl rubber, EPR, SBR, and other elastomers are summarized in Table 14.2. Since most commercial elastomers contain varying amounts of additives, fillers, and crosslinking agents, which can drastically alter their properties, the values in Table 14.2 are intended only for comparative purposes. Some of the properties listed that are important in elastomeric applications have not been considered to any significant extent in previous discussions. For example, a lower use temperature (as well as an upper use temperature) is important in elastomeric applications. Unlike other amorphous polymers, elastomers are utilized above their glass transition temperatures. If an application requires exposure to very low temperatures, then an elastomer with a low $T_g$ is required. Otherwise, as the ambient temperature approached the glass transition temperature, the material would begin to become glassy and lose its desirable elastomeric properties.

Another property that can be important in elastomeric applications is damping capacity. This is a measure of the ability of a material to absorb vibrational energy as opposed to transmitting it. It is a viscoelastic phenomenon involving stress–strain hysteresis (Figure 14.1). Elastomers differ in their ability to do this, and their applications often depend on this ability (or lack of it).

Since vibration problems occur in many ways in automobiles, automobiles are an important example of a product whose design involves material damping. For example, a primary function of the engine mounts is to reduce the vibration of the engine that would otherwise be transmitted to the frame of the car. The elastomer in the engine mount internally converts much of the vibrational strain energy into heat. This represents energy that cannot be transmitted to the frame as vibration.* Butyl rubber is a good choice for this application. On the other hand, using the same elastomer in tires would be disastrous. Because heat is one of the greatest enemies of a tire, a low-damping elastomer is preferred over a high-damping one [5]. In truck and heavy equipment tires, for example, natural rubber is a preferred material because of its relatively low damping qualities.

## THERMOPLASTIC ELASTOMERS

Thermoplastic elastomers are an exception to the crosslink requirement in elastomers. An example of these is a styrene butadiene block copolymer in which each molecule consists of two blocks of styrene repeating units separated by a block of butadiene repeating units:

*The actual mechanisms are complex. References [6] and [7] contain further information.

**TABLE 14.1 Typical Structures and Applications of Commercial Elastomers**

| Polymer | Repeating Unit | Crosslinking Agent | Applications |
|---|---|---|---|
| Polyisoprene | | Sulfur | Automobile tires |
| SBR | A random copolymer of styrene and butadiene units | Sulfur or peroxide | Automobile tires |
| Neoprene | | Zinc and magnesium oxide | Cable sheathing; hoses; weather stripping |
| Butyl | A random copolymer of isobutene units and a small number of isoprene units | Sulfur (through isoprene units) | Inner tubes; engine mounts and other high damping applications |

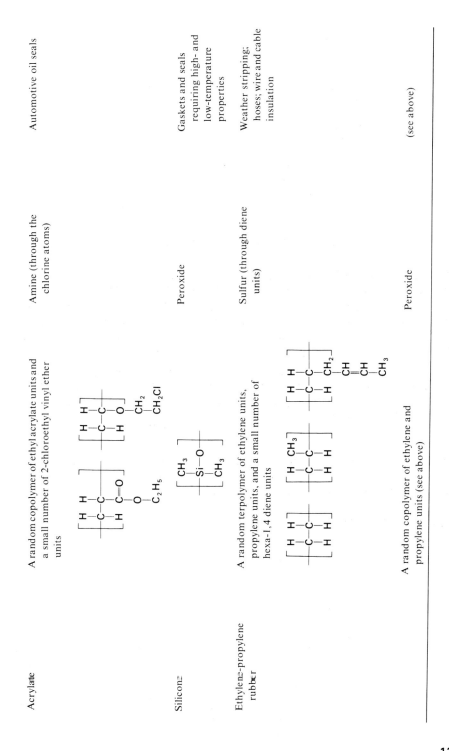

| Acrylate | A random copolymer of ethyl acrylate units and a small number of 2-chloroethyl vinyl ether units | Amine (through the chlorine atoms) | Automotive oil seals |
| Silicone | | Peroxide | Gaskets and seals requiring high- and low-temperature properties |
| Ethylene-propylene rubber | A random terpolymer of ethylene units, propylene units, and a small number of hexa-1,4 diene units | Sulfur (through diene units) | Weather stripping; hoses; wire and cable insulation |
| | A random copolymer of ethylene and propylene units (see above) | Peroxide | (see above) |

**TABLE 14.2 Typical Properties of Commercial Elastomers**

| Polymer | Tensile Properties | | | Damping Capacity | Lower Use Temperature | Upper Use Temperature | Gas Permeability | Tear Resistance | Hydrocarbon Resistance |
| | Strength | Modulus at 300–400% Elongation | Elongation | | | | | | |
|---|---|---|---|---|---|---|---|---|---|
| Natural rubber | 4500 psi | 2500 psi | 600% | Low | −60°C | +100°C | High | Good | Poor |
| SBR (normal) | 3000 psi | 2000 psi | 500% | Fairly high | −55°C | +100°C | High | Poor | Poor |
| (cold) | 3800 psi | 2500 psi | 550% | Fairly low | −55°C | +100°C | High | Fair | Poor |
| Neoprene | 4000 psi | 1000 psi | 800% | Fairly high | −45°C | +80–100°C | Moderate | Good | Good (aliphatics) Fair (aromatics) |
| Butyl | 3000 psi | 1000 psi | 400% | High | −50°C | +120°C | Low | Excellent | Poor |
| Acrylate | 2500 psi | — | 400% | Fairly low | −25°C | +200°C | Low | Good | Good |
| Silicone | 1500 psi | — | 600% | High | −90°C | +250°C | — | Poor | Poor |
| EPR | 3000 psi | — | — | Fairly high | −60°C | +150°C | — | Poor | Poor |

(*Adapted from* [1–4].)

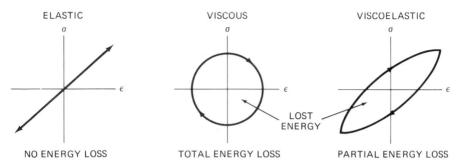

Figure 14.1 Damping in Elastic, Viscous, and Viscoelastic Materials.

This is in direct contrast to styrene butadiene rubber (SBR), which is a chemically crosslinked, elastomeric random copolymer of styrene and butadiene units. It is because of the block nature of the styrene butadiene thermoplastic elastomer that the crosslinking requirement can be circumvented. The styrene segments are not compatible with the butadiene segments, so they tend to segregate in the bulk material. The result is a blend of glassy, brittle, polystyrenelike regions separated by (yet covalently bonded to) soft, rubbery, polybutadienelike regions. The glassy regions act as crosslinks for the butadiene segments.

This material has found applications in a variety of products, including shoe soles, sealants, adhesives, and coatings. A major advantage, of course, is that it can be softened, shaped, and cooled repeatedly, in contrast to other elastomers, which are thermosets. The softening process required to make a thermoplastic elastomer flow involves heating the material above the glass transition temperature of the polystyrene regions. Solidification occurs when the material cools and the glassy, polystyrenelike regions form. Since no degradation occurs in doing this, the scrap is recyclable.

## PROBLEM SET

1. What are the four basic mechanical properties characteristic of elastomers?
2. What conditions usually exist when the properties in Problem 1 are observed?
3. Which of the polymers listed in Table 3.2 are often utilized as elastomers?
4. To what transition does the reported lower use temperature in Table 14.2 relate?
5. Write a short paragraph describing what happens on a molecular level when an elastomer is stretched and released.
6. Uncrosslinked, amorphous polymers above their $T_g$ can be somewhat rubbery. Explain why this can occur.
7. What characteristics of PVC prevent it from being elastomeric at room temperature?
8. Repeat Problem 7 for:
   (a) Polystyrene
   (b) Polypropylene
   (c) Nylon

**9.** The following graph shows the general effects of the amount of crosslinking agent on the elongation to break for a rubbery material.

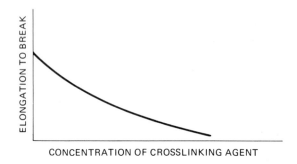

CONCENTRATION OF CROSSLINKING AGENT

Explain this trend.

**10.** Draw a graph of modulus versus amount of crosslinking agent.

**11.** Explain how thermoplastic elastomers circumvent the crosslink requirement for elastomers.

**12.** What advantage does a thermoplastic elastomer have over other elastomers?

**13.** Why does neoprene have better hydrocarbon solvent resistance than natural rubber (Table 14.2)?

## REFERENCES

[1] Billmeyer, F. W., *Textbook of Polymer Science*, Interscience, New York (1962).

[2] Rodriguez, F., *Principles of Polymer Systems*, McGraw-Hill, New York (1970).

[3] Morton, M. (ed.), *Introduction to Rubber Technology*, Rheinhold, New York (1959).

[4] Whitby, G. S., C. C. Davis, and R. F. Dunbrook, (eds.), *Synthetic Rubber*, Wiley, New York (1954).

[5] Rosen, S. L., *Fundamental Principles of Polymeric Materials*, Wiley, New York (1981).

[6] Woodward, A. E., and J. A. Sauer, Mechanical Relaxation Phenomena, in *Physics and Chemistry of the Organic Solid State* (D. Fox et al., eds.), Wiley-Interscience, New York (1965).

[7] Ferry, J. D., *Viscoelastic Properties of Polymers*, Wiley, New York (1970).

# electrical properties

## chapter 15

If there is one area in which polymers are largely indispensable, it is probably in electrial and electronic applications. Removing polymers from products such as electrical motors, electrical generators, computers, television, and radio would eliminate these items as we know them and would completely change our way of living. While it is true that ceramics and glasses could serve as insulators in some of these areas, these applications would be relatively cumbersome and limited.

The electrical uses of polymers involve a variety of directly applicable properties. Some of these are

1. Volume resistivity
2. Dielectric strength
3. Arc resistance
4. Dielectric constant
5. Dissipation factor

These and other critical electrical factors must be evaluated in electrical applications. The exact requirements depend on the voltage, power, frequency of operation, temperature, time in service, environment, etc.

It is important to recognize, however, that in addition to these properties, polymers have critical nonelectrical properties in electrical applications. Many of these characteristics are discussed elsewhere in the text; and in some cases the reader can anticipate the applicability of these to electrical applications. For example, a certain degree of flame resistance is generally required in an insulating coating for a conductor, in addition to specific electrical properties. This chapter introduces only the directly applicable electrical properties cited above.

## VOLUME RESISTIVITY

Volume resistivity is the resistance that a material presents to the flow of electrical current (charge/s), which in most cases involves electrons as opposed to ions or other charges. ASTM D 257 [1] is the recommended procedure for determining this property for polymers [1]. The units of resistivity are ohm-cm, or resistance (ohms) per unit length (cm) of material per unit area of cross section ($cm^2$). Common values vary widely from that of silver ($1.6 \times 10^{-6}$ ohm-cm) to that of paraffin ($\sim 10^{19}$ ohm-cm). Table 15.1 lists some typical values for a variety of materials.

For the most part, polymers can be considered to be insulators; that is, their resistance to current flow is rather high. This directly reflects the basic nature of polymers as described in previous chapters, in that there are essentially no "free" electrons as there are in metals. Except for impurities, additives, and fillers, the electrons are tied up in the covalent bonds and are therefore not available for movement throughout the material. Thus polymers typically have resistivities millions of times larger than those of metals.*

TABLE 15.1  Approximate Volume Resistivity
Values of Some Common Polymers
and Other Materials

| Material | Volume Resistivity (ohm-cm) at 23° C |
|---|---|
| Silver | $1.6 \times 10^{-6}$ |
| Copper | $2 \times 10^{-6}$ |
| Aluminum | $3 \times 10^{-6}$ |
| Steel | $2 \times 10^{-5}$ |
| Phenol formaldehyde | $10^{10}-10^{12}$ |
| Epoxy | $10^{12}-10^{15}$ |
| Polyethylene | $10^{15}-10^{16}$ |
| Polystyrene | $10^{18}$ |
| Paraffin | $10^{19}$ |

(*Adapted from* [2] *and* [3].)

## DIELECTRIC STRENGTH

Dielectric breakdown voltage is the voltage that a given insulator can withstand before electrical breakdown occurs. Dielectric strength is the ratio of this breakdown voltage to the thickness of the material. The units are volts per centimeter. This property is typically measured in accordance with ASTM D 149 [1]. In this test electrodes are placed on opposite sides of the material. The AC

---

*Although some special conductive polymers (polyacetylene, polypyrrole, polyparaphenyl-ene, etc.) have been produced, they are still in the research stage.

TABLE 15.2  Typical Dielectric Strengths of Some Common
Polymers at 60 Hz (Short-Term)

| Polymer | Dielectric Strength $\times$ $10^{-3}$ (volts/cm) |
|---|---|
| Low-density polyethylene | 180–390 |
| Polystyrene | 200–280 |
| Nylon 6/6 | 240 |
| Polyphenylene oxide | 160–220 |
| High-density polyethylene | 190–200 |
| Epoxy | 160–200 |
| Polytetrafluoroethylene | 160–200 |
| Nylon 6 | 160 |
| Polycarbonate | 160 |

(*Adapted from* [3].)

voltage across the electrodes is increased either at a uniform rate or in steps until breakdown occurs. Generally, the observed breakdown voltage is higher for rapid increases in voltage than for stepped increases or slower rates of voltage rise.

Table 15.2 lists typical values for some common polymers at 60 Hz. It is important to note that this property is sensitive to frequency, time, and moisture. Dielectric strength is extremely important not only in high-voltage applications but also in situations in which weight and size are limited, such as in the windings of motors.

## ARC RESISTANCE

Arc resistance is a measure of the tendency for electrical breakdown to occur along an insulating surface via formation of a conductive path. Often this path is a carbon track resulting from the severe degradation of polymer molecules. Once a continuous path is formed, it will trigger an electrical failure. A common example of a situation where this problem occurs is in the distributor caps of automobile engines. The presence of dirt and/or moisture, of course, aggravates the situation.

Measurement of the arc resistance of clean, dry polymers is specified in ASTM D 495 [1]. In this test two electrodes are held on the surface of the material to be tested, 0.635 cm apart. A high-voltage, low-current arc is then applied close to the surface of the material according to a specified time schedule, and the time required for a conducting path to form is determined. Table 15.3 lists typical values for some common polymers. In general, aromatic-based polymers such as phenol formaldehyde tend to track more readily than many other polymers. Fillers (such as alumina trihydrate) have been found to increase arc and track resistance greatly [2].

TABLE 15.3 Arc Resistance of Some
Common Polymers

| Polymer | Arc Resistance (s) |
| --- | --- |
| Polytetrafluoroethylene | >200 |
| Polypropylene | 136–185 |
| Low-density polyethylene | 135–160 |
| Nylon 6/6 | 130–140 |
| Polystyrene | 60–140 |
| Epoxy | 45–120 |
| Polycarbonate | 10–120 |
| Polyphenylene oxide | 75 |

(*Adapted from* [3].)

## DIELECTRIC CONSTANT

Dielectric constant can be viewed as the ability of a material, when placed between the plates of a capacitor, to cause an increase in the capacitance over that when the plates are separated by vacuum. It is expressed as the ratio of the capacitances:

$$\text{Dielectric Constant} = \frac{\text{Capacitance with the Material in Question}}{\text{Capacitance with Vacuum between the Plates}}$$

A high dielectric constant is frequently desirable, since it permits the use of smaller capacitors in circuits. On the other hand, a low dielectric constant is preferred in a material used to insulate one circuit component from another. Typically it is strongly dependent on both frequency and temperature. Values at radar frequencies (approximately 10 GHz) can be much different from values at common power-distribution frequencies (60 Hz). The measurement of this property is specified in ASTM D 150 [1]. Some typical values for various materials at 60 Hz and 23°C are listed in Table 15.4.

TABLE 15.4 Typical Dielectric Constants of Some
Common Polymers

| Polymer | Dielectric Constant (60 Hz; 23°C) |
| --- | --- |
| Polytetrafluoroethylene | 2.1 |
| Polypropylene | 2.2 |
| Low-density polyethylene | 2.3 |
| High-density polyethylene | 2.3 |
| ABS | 3.2 |
| Polycarbonate | 3.2 |
| Polyformaldehyde | 3.7 |
| Nylon 6/6 (dry) | 3.8 |
| Nylon 6/6 (50% relative humidity) | 8.2 |

(*Adapted from* [4].)

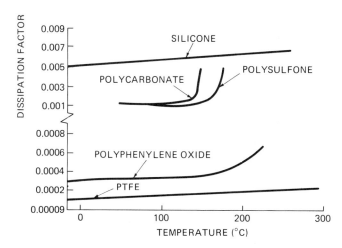

**Figure 15.1** Dissipation Factor as a Function of Temperature for Several Polymers. (*Adapted from* [2].)

## DISSIPATION FACTOR

Dissipation factor is a measure of the heat generated in an insulating material subjected to an oscillating electric field. It is generally preferred that this value be small, to minimize the heating of the material and the effect of this heat on other portions of the network [1]. This heat generation is the result of molecular motion in response to the electric field, and in this regard is similar to the mechanical damping (see Chapter 14) that occurs in materials subjected to an oscillating mechanical stress field. In both cases, a hysteresis effect occurs, and energy is lost. As was the case for dielectric constant, the dissipation factor varies with both frequency and temperature. Figure 15.1 illustrates the effect of temperature on the dissipation factors of several polymers. The sensitivity of this property to $T_g$ and $T_m$ is readily evident.

**TABLE 15.5  Typical Dissipation Factors for Some Common Polymers**

| Polymer | Dissipation Factor (60 Hz; 23° C) |
|---|---|
| Polytetrafluoroethylene | 0.0005 |
| Polypropylene | 0.0001 |
| Low-density polyethylene | 0.0001 |
| High-density polyethylene | 0.0001 |
| ABS | 0.0065 |
| Polycarbonate | 0.0009 |
| Polyformaldehyde | 0.0048 |
| Nylon 6/6 (dry) | 0.009 |
| Nylon 6/6 (50% relative humidity) | 0.14 |

(*Adapted from* [4].)

ASTM D 150 [1] specifies the procedure for measuring dissipation factor. Some typical values for common polymers at 60 Hz and 23°C are listed in Table 15.5.

## PROBLEM SET

1. List 12 products that do not utilize electricity directly in either their use or their manufacture.

2. Distinguish between conductors and insulators. Explain why metals typically fall into the former category while polymers typically are included in the latter.

3. How is the resistivity of a material related to the resistance of an object of given dimensions?

4. Given a polyethylene rod (volume resistivity $= 5 \times 10^{15}$ ohm-cm) that is 1.588 mm in diameter and 1 m long, how long would a solid copper wire of the same diameter have to be in order to have the same resistance?

5. Distinguish between dielectric constant and dielectric strength. What are some applications in which each is important?

6. If you wished to increase the dielectric constant of a polymer by adding a filler, what type of inorganic material would probably be a very efficient choice?

7. What properties (both electrical and nonelectrical) would probably be important in the following products?
   (a) A distributor cap for a gasoline automobile engine.
   (b) A high-voltage trailing power cable for a mine car.
   (c) The insulation for the wiring in a house.

## REFERENCES

[1] American Society for Testing and Materials, *ASTM Standards, Part 35*, Philadelphia (1981).

[2] Harper, C. A., Plastics: Today's Choice in E/E Part Design, *Plastics Design Forum* (April 1981), pp. 15–54.

[3] Brandrup, J., and E. H. Immergut, *Polymer Handbook*, 2nd ed., Wiley, New York (1975).

[4] *Modern Plastics Encyclopedia*, vol. 55, no. 10A, McGraw-Hill, New York (1978–1979).

# processing of polymers

# polymerization mechanisms

## chapter 16

## CHAIN AND STEPWISE POLYMERIZATION

It is not the purpose of this book to teach the details of polymerization processes. There are a number of polymer texts that examine this in detail [1–4]. However, the following discussion should provide a general appreciation of what occurs during polymerization, as a basis for other topics dealt with in this book.

Most polymerization processes fall into one of two categories: chain polymerization and stepwise polymerization. These differ primarily in the way in which the molecules grow. Assume that there are two reactors:

one filled with a monomer that polymerizes by a chain mechanism and one filled with a monomer that polymerizes in a stepwise fashion. If polymerization was initiated in both at time $t_0$, and terminated at time $t_1$, an analysis of the contents of each would show completely different products. In the chain polymerization reactor one would find some monomer and some polymer of high molecular weight, but very little material of intermediate molecular weight. In the stepwise reactor one would find a distribution of material of intermediate molecular weight and little monomer or polymer of high molecular weight.

The reason for these differences is the way in which the molecules form. In chain polymerization a reactive site is created that attacks monomer in a rapid fashion and forms a complete polymer chain almost instantly:

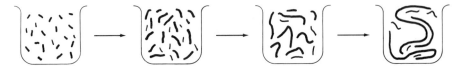

In stepwise polymerization, on the other hand, the molecular length increases throughout the reactor in steps or blocks as first two units join, and then these dimers unite to form a four-unit structure, and so on:

Monomers that polymerize by a chain mechanism are typically those in which a carbon–carbon double bond is reactive. Ethylene, styrene, propylene, tetrafluoroethylene, methyl methacrylate, and the various dienes (isoprene, butadiene, etc.) are examples of this. Stepwise polymerization, on the other hand, usually involves reactive groups such as hydroxyls (—OH), carboxyls (—$\overset{\overset{\textstyle O}{\|}}{C}$—OH), and hydrogen atoms attached to nitrogen. The nylons, polyesters, urethanes, epoxies, and phenolics are examples of polymers that usually form by this general process.

More detailed examples of these processes are given in the following section.

### Chain Polymerization Examples

The first step in chain polymerization is initiation, and this usually involves the formation of free radicals.* These free radicals can be derived from initiators such as azobisisobutyronitrile,

$$(CH_3)_2-C-N{=}N-C-(CH_3)_2$$
$$\underset{\textstyle C{\equiv}N}{|} \qquad \underset{\textstyle C{\equiv}N}{|}$$

and benzoyl peroxide.

Both of these compounds are readily decomposed by heat and form free radicals as shown:

*A free radical is a reactive, electrically neutral molecule. Its reactivity stems from the presence of an unpaired electron.

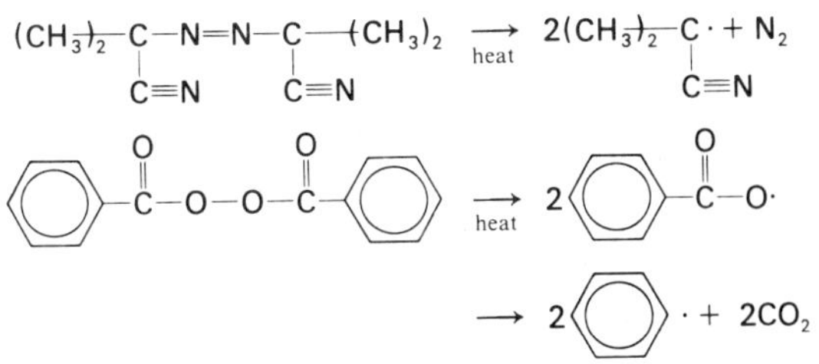

$$(CH_3)_2-C-N=N-C-(CH_3)_2 \xrightarrow{\text{heat}} 2(CH_3)_2-C\cdot + N_2$$

with $C\equiv N$ groups shown below each carbon, and $C\equiv N$ on the product.

$$\text{Ph}-\overset{O}{\underset{\|}{C}}-O-O-\overset{O}{\underset{\|}{C}}-\text{Ph} \xrightarrow{\text{heat}} 2\,\text{Ph}-\overset{O}{\underset{\|}{C}}-O\cdot$$

$$\longrightarrow 2\,\text{Ph}\cdot + 2CO_2$$

The free radicals are the components containing the "·," which symbolizes an unpaired electron. For simplicity the free radicals that are formed can be represented as "R ·."

These free radicals react with the double bonds in the monomer molecules. This reaction, which is called propagation, proceeds as follows:

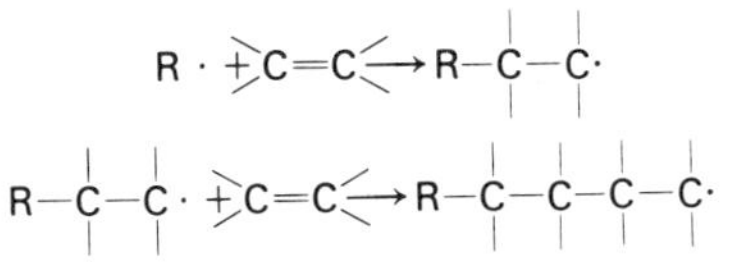

$$R\cdot + C=C \longrightarrow R-C-C\cdot$$

$$R-C-C\cdot + C=C \longrightarrow R-C-C-C-C\cdot$$

In general:

$$R\left[\begin{array}{c}C-C\end{array}\right]_n C-C\cdot + C=C \longrightarrow R\left[\begin{array}{c}C-C\end{array}\right]_{n+1} C-C\cdot$$

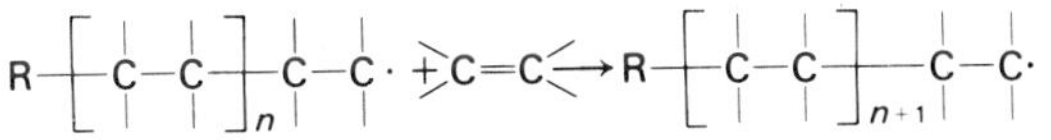

This occurs rapidly, as was indicated at the beginning of the chapter. A completed molecule can be formed in a matter of seconds [2].

The final step in chain polymerization is the cessation of chain growth, or termination. In chain polymerization initiated by free radicals, for example, this can occur in one or more of the following ways:

1. Combination of two growing chains:

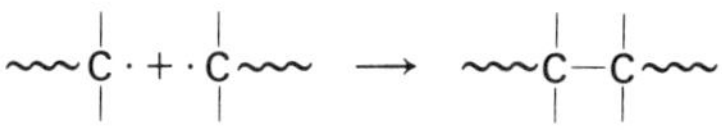

$$\sim\sim C\cdot + \cdot C\sim\sim \longrightarrow \sim\sim C-C\sim\sim$$

2. Disproportionation:

$$\sim\sim\overset{H\ H}{\underset{H\ H}{C-C}}\cdot + \cdot\overset{H\ H}{\underset{H\ H}{C-C}}\sim\sim \longrightarrow \sim\sim\overset{H}{C}=C\overset{H}{\underset{H}{}} + H-\overset{H\ H}{\underset{H\ H}{C-C}}\sim\sim$$

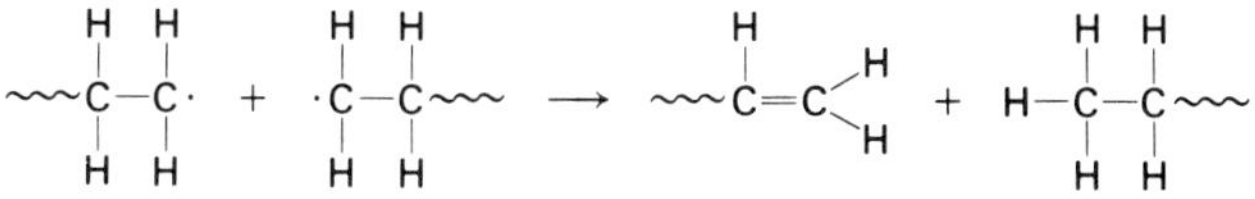

3. Transfer to dead polymer (branching will occur here):

The mechanism that predominates in any given case depends on the type of monomer, its purity, and the presence of additives such as chain transfer agents.

The actual termination methods are not of great significance as far as this text is concerned. What the student should recognize and retain is that the final length of any particular polymer molecule is determined largely by probability. This is consistent with the fact that synthetic polymers tend to be polydisperse; that is, the individual molecules have different lengths.

While chain polymerization is sometimes called free-radical polymerization, one must remember that chain polymerization is a general method of chain growth rather than a mechanism of propagation. There are several other types of chain polymerization, which involve anions and cations rather than free radicals. These include the cationic, anionic, and heterogeneous stereospecific polymerization processes. These are very important in producing certain polymers. For example, cationic polymerization is useful in the low-temperature commercial polymerization of polyisobutylene,

$$\left[\begin{array}{cc} H & CH_3 \\ | & | \\ C - C \\ | & | \\ H & CH_3 \end{array}\right]_n$$

Anionic polymerization is useful for the production of styrene–butadiene block copolymers (Chapter 14) and is the only known scientific technique for producing nearly monodisperse synthetic polymers. Heterogeneous stereospecific processes are used to produce stereoregular polypropylene and linear (high-density) polyethylene. The interested reader is urged to consult the organic polymer chemistry texts listed at the end of this chapter for additional information concerning these processes.

### Stepwise Polymerization Examples

Stepwise polymerization mechanisms can involve many types of processes, including oxidative coupling reactions and aromatic substitution. However, the majority of polymers produced by stepwise mechanisms utilize the reaction of dissimilar functional groups. These groups include:

1. Carboxyl groups:

$$\sim\sim\sim\overset{\displaystyle O}{\overset{\displaystyle \|}{C}}-OH$$

2. Hydroxyl groups:

$$\sim\sim\sim OH$$

3. Hydrogen atoms attached to nitrogen:

$$\sim\!\!\sim\!\!\overset{\displaystyle\{}{\underset{}{N}}\!\!-\!\!H$$

4. Epoxide groups:

$$\sim\!\!\sim\!\!\overset{\displaystyle O}{\underset{\underset{\displaystyle H\ \ H}{|\ \ \ |}}{C-C}}\!\!-\!\!H$$

5. Isocyanate groups:

$$\sim\!\!\sim\!\!N\!=\!C\!=\!O$$

In the manufacture of polyethylene terephthalate (a linear polyester), for example, one usually starts with dimethyl terephthalate,

$$CH_3O\!-\!\overset{O}{\overset{\|}{C}}\!-\!\!\bigcirc\!\!-\!\overset{O}{\overset{\|}{C}}\!-\!OCH_3$$

and ethylene glycol (the main constituent of antifreeze),

$$HO\!-\!\overset{\overset{\displaystyle H\ \ H}{|\ \ \ |}}{\underset{\underset{\displaystyle H\ \ H}{|\ \ \ |}}{C-C}}\!-\!OH$$

When these are mixed, the end groups on the ester react with the hydroxyl groups on the glycol in the following manner:

$$CH_3O\!-\!\overset{O}{\overset{\|}{C}}\!-\!\!\bigcirc\!\!-\!\overset{O}{\overset{\|}{C}}\!-\!OCH_3 \ + \ HO\!-\!\overset{\overset{\displaystyle H\ \ H}{|\ \ \ |}}{\underset{\underset{\displaystyle H\ \ H}{|\ \ \ |}}{C-C}}\!-\!OH$$

$$\longrightarrow \ CH_3O\!-\!\overset{O}{\overset{\|}{C}}\!-\!\!\bigcirc\!\!-\!\overset{O}{\overset{\|}{C}}\!-\!O\!-\!\overset{\overset{\displaystyle H\ \ H}{|\ \ \ |}}{\underset{\underset{\displaystyle H\ \ H}{|\ \ \ |}}{C-C}}\!-\!OH \ + \ CH_3OH$$

The $-\overset{O}{\overset{\|}{C}}-O-$ linkage formed with the loss of methanol is called the polyester linkage and is the common factor in all polyesters (what exists between these groups is responsible for the differences among polyesters).

The compound formed in this initial reaction also has reactive groups and can therefore continue to increase in molecular weight. As each subsequent reaction occurs, the molecular weight will increase in discrete steps. In general terms, the stepwise polymerization of $n$ molecules of dimethyl terephthalate and $n$ molecules of ethylene glycol can be written as

The "$-1$" in the $(2n - 1)$ occurs because the finished polymer chain has one unreacted group on each of its ends.

In addition to linear polyesters such as polyethylene terephthalate, many other important polymers are produced by stepwise mechanisms. The general reactions for those listed in Table 3.2 are summarized below.

## Polycarbonate

## Polyurethanes

*Note*: This is an example of a stepwise reaction with no byproduct. Crosslinked polyurethane systems are produced with alcohols having higher functionalities.

## Phenol formaldehyde

(phenol)            (formaldehyde)
f = 3*                  f = 2

## Urea formaldehyde

(urea)              (formaldehyde)
f = 4*                  f = 2

## Nylon 11**

$\omega$-aminoundecanoic acid
f = 2

## Nylon 6/6**

(hexamethylenediamine)       (adipic acid)
f = 2                         f = 2

*This is the maximum value. In practice the average functionality will be lower.

**The numerals (6, 10, 11) in the nylon nomenclature refer to the number of carbon atoms in the monomer(s).

## Nylon 6/10**

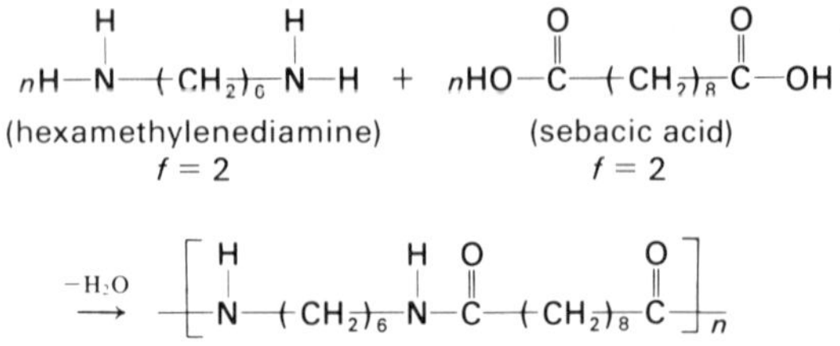

$$nH-\overset{H}{\underset{|}{N}}-(CH_2)_6-\overset{H}{\underset{|}{N}}-H \;+\; nHO-\overset{O}{\overset{\|}{C}}-(CH_2)_8-\overset{O}{\overset{\|}{C}}-OH$$

(hexamethylenediamine)          (sebacic acid)

$f = 2$                     $f = 2$

$$\xrightarrow{-H_2O} \left[ -N-(CH_2)_6-\overset{H}{\underset{|}{N}}-\overset{O}{\overset{\|}{C}}-(CH_2)_8-\overset{O}{\overset{\|}{C}} \right]_n$$

## Melamine formaldehyde

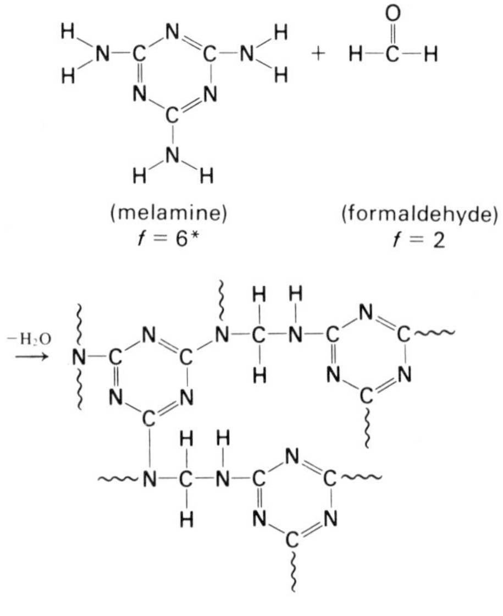

(melamine)           (formaldehyde)

$f = 6*$               $f = 2$

## Polyphenylene oxide

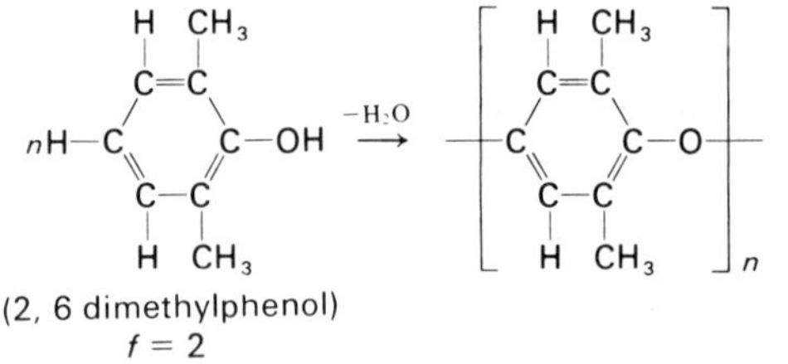

(2, 6 dimethylphenol)

$f = 2$

*This is the maximum value. In practice the average functionality will be lower.

**The numerals (6, 10, 11) in the nylon nomenclature refer to the number of carbon atoms in the monomer(s).

## Polyarylsulfone

(diphenylene oxide sulfonyl chloride)
f = 2

The reader is encouraged to consult an organic polymer chemistry text for detailed discussions of the polymerization of these and other polymers.

A question that frequently arises when discussing many stepwise polymerization products is whether the polymers are random copolymers. For example, should polyethylene terephthalate be considered a random copolymer of

and

units? It is customary not to consider them as such. The criterion is that in a copolymer, each of the different types of monomer units should be capable of forming a homopolymer. In an ethylene–propylene copolymer, for example, this is true; both polyethylene and polypropylene exist. It is not the case with the stepwise polymers discussed here, however; and they should therefore be considered homopolymers.

## STOICHIOMETRY IN STEPWISE POLYMERIZATION

In stepwise reactions one is often faced with the prospect of mixing two components together. For example, in the production of nylon 6/6, the same number of diamine and diacid molecules is desirable so that the chains can grow to a large size. If too much of one component exists, the resulting molecular weight is too low. It is critical in many cases, therefore, to use the correct amount of each component. Commercially, this is done for nylon 6/6 by first producing a salt:

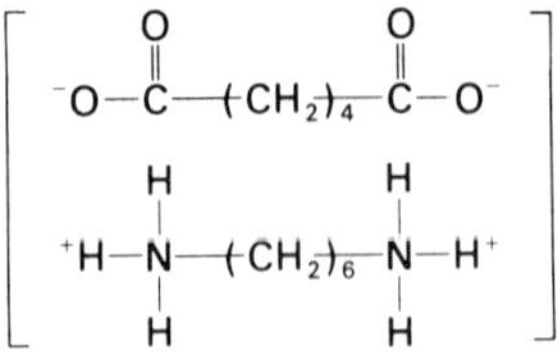

$$
\begin{bmatrix}
\begin{array}{c}
\overset{O}{\overset{\|}{\phantom{|}}}\quad\overset{O}{\overset{\|}{\phantom{|}}} \\
{}^{-}O-C-\!\!\!\!(CH_2)_4\!\!-\!\!C-O^{-} \\[2mm]
H \qquad\quad H \\
{}^{+}H-N-\!\!\!\!(CH_2)_6\!\!-\!\!N-H^{+} \\[1mm]
H \qquad\quad H
\end{array}
\end{bmatrix}
$$

which guarantees a 1 : 1 molar ratio of the two components. This technique is also used to produce nylon 6/10. In some cases, such as epoxies, however, this approach is not used. The following example of the curing of an epoxy resin illustrates how stoichiometric mixtures can be determined in these situations.

Many common epoxy resins are based on diglycidyl ether of bisphenol A (DGEBA) and have the following general structure:

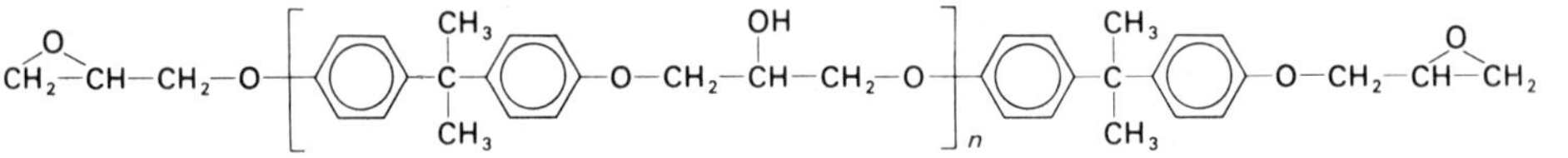

They are called epoxies because of the epoxide groups,

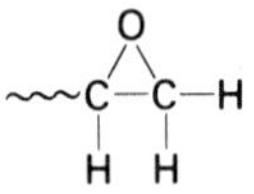

on the ends of the molecules. In its simplest form (DGEBA), $n = 0$, and the molecular weight is 340 g/mol. Typical formulations have average molecular weights higher than this, which indicates that some of the molecules have $n > 0$.

Epoxies can be polymerized, or "cured," in a number of ways, including the use of amines, acid anhydrides, and ionic catalysts. The amine curing agent is very common, and a typical member of this group is diethylenetriamine (DTA):

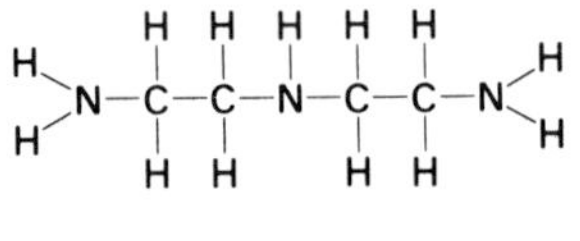

In this molecule the hydrogens that are attached to the nitrogens are reactive, and they react with the epoxide groups in the following manner:

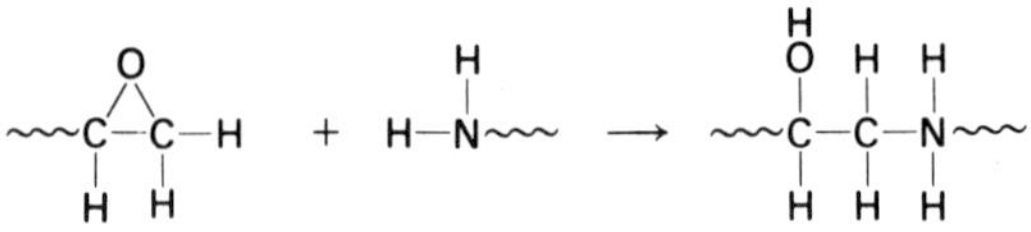

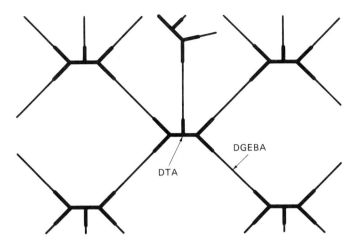

**Figure 16.1** Network Formed by the Reaction of DGEBA and DTA (Stoichiometric).

Ideally, all of the reactive groups on the amine and the epoxy will react, creating a crosslinked polymer as illustrated in Figure 16.1. In this diagram the fine black lines represent the reacted DGEBA molecules and the heavier-lined structures represent the reacted DTA molecules. If too much of one component exists in the reactive mixture initially, the structure which results will differ from that in Figure 16.1. Specifically, if an excess of amine was added, then a structure similar to that in Figure 16.2 could result from polymerization. All of the reactive groups on the amine molecules cannot react in this case, since there are too few DGEBA

**Figure 16.2** Network Formed by the Reaction of DGEBA and DTA (Nonstoichiometric—Excess of DTA).

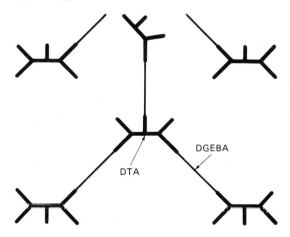

molecules. The network that results lacks the continuity present in the stoichiometric structure.

In order for all of the reactive groups (on both the amine and the epoxy) to react, five DGEBA molecules must be present for every two DTA molecules. A stoichiometric mixture will result, therefore, when the molar ratio of DGEBA molecules to DTA molecules is $5:2$. In terms of the mass ratio of reactants, this molar ratio is equivalent to

$$\frac{5 \text{ moles DGEBA}}{2 \text{ moles DTA}} = \frac{(5 \text{ moles DGEBA}) (340 \text{ g/mol})}{(2 \text{ moles DTA}) (103 \text{ g/mol})} = \frac{1700 \text{ g DGEBA}}{206 \text{ g DTA}}$$

The amount of DTA required is typically expressed in *phr*, or parts per hundred parts by weight of resin (DGEBA). The above ratio may be converted to phr using the following relationship:

$$\frac{1700 \text{ DGEBA}}{206 \text{ DTA}} = \frac{100 \text{ DGEBA}}{X \text{ DTA}}$$

where $X$ is the phr (12.1 in this example). Calculations for other systems are similar.

## PROBLEM SET

1. What are the differences between chain and stepwise polymerization as regards:
   (a) The type of monomer generally involved?
   (b) The way in which long chains are formed?
2. Which of the polymers in Table 3.2 are produced by chain polymerization mechanisms?
3. (a) Write the *general* equation for the polymerization of $n$ moles of ethylene glycol with $n$ moles of dimethyl terephthalate:

   (b) What are the functionalities of the alcohol and the ester?
   (c) What is the technical name for the polymer?
   (d) Write the first four steps of the polymerization process in part (a), confirming the $(2n - 1)$ factor.
   (e) What would happen to the process if one half of the ethylene glycol was replaced with ethanol?
4. Explain why most synthetic polymers are polydisperse.
5. What are some examples of monodisperse polymers?
6. Monomers possess different tendencies to react with other monomers and with

themselves. What tendencies to react would have to exist for monomer $A$ and monomer $B$ in order to obtain the following via a chain polymerization mechanism?
(a) A random copolymer ($ABAABABB$)
(b) An alternating copolymer ($ABABABAB$)
(c) A block copolymer ($AAAABBBB$)

7. Following the example in the text, calculate the required phr for DGEBA for the following curing agents:
(a) Triethylenetetramine (TETA)

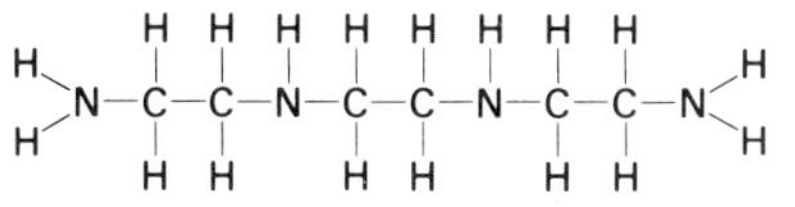

(b) $m$-Phenylenediamine

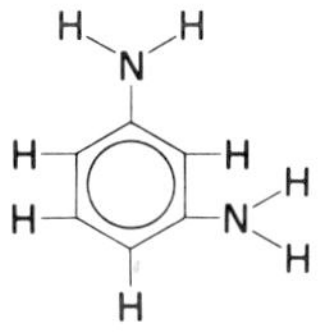

8. Using nylon 6/6 as an example, explain how an excess of one reactant could result in low-molecular-weight product during stepwise polymerization.

9. How many grams of $m$–phenylenediamine are required to cure a 200-gram batch of an epoxy resin ($f = 4$; molecular weight $= 500$ g/mol)?

10. The commercial production of polyethylene terephthalate utilizes the dimethyl ester of the acid,

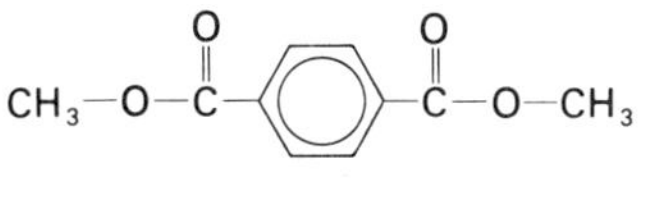

rather than the acid,

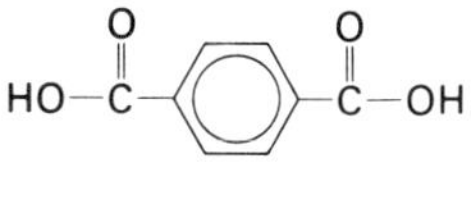

Explain why.

**Answer:**

In stepwise polymerization it is important that the reactants be relatively free of impurities. This is necessary in order to obtain high molecular weights. Terephthalic acid is difficult to purify, whereas the ester is not, so the latter is preferred. The resulting repeating unit is the same whether the ester or the acid is used. In the ester case methanol is the byproduct of polymerization; in the acid case it is water.

## REFERENCES

[1] Flory, P. J., *Principles of Polymer Chemistry*, Cornell University Press, Ithaca, N.Y. (1953).

[2] Saunders, K. J., *Organic Polymer Chemistry*, Chapman and Hall, London (1973).

[3] Odian, G., *Principles of Polymerization*, McGraw-Hill, New York (1970).

[4] Seymour, R. B., and C. E. Carraher, Jr., *Polymer Chemistry: An Introduction*, Dekker, New York (1981).

The polymerization mechanisms outlined in Chapter 16 are generally carried out industrially by one or more of four basic processes: bulk, solution, emulsion, and suspension polymerization. In each case a critical point is the control of heat generated during polymerization. A discussion of the source of this heat is given below, followed by brief descriptions of each of these industrial polymerization methods.

## HEAT GENERATION DURING POLYMERIZATION

Heat is released in both chain and stepwise polymerization. As an example, consider the chain polymerization of one mole of a vinyl monomer:

$$n \begin{matrix} H \\ H \end{matrix} C=C \begin{matrix} H \\ R \end{matrix} \longrightarrow \left( \begin{matrix} H & H \\ | & | \\ C-C \\ | & | \\ H & R \end{matrix} \right)_n$$

$$R = H, CH_3, Cl, \text{ etc.}$$

$$n = 6.02 \times 10^{23}$$

For heat calculation purposes, this is equivalent to adding one monomer unit to a growing polymer chain $6.02 \times 10^{23}$ times. To calculate the amount of heat generated during this process, the strength of a carbon–carbon double bond and the strength of a carbon–carbon single bond are needed, since these values are the amount of energy required to break these bonds and the amount of energy released when these bonds are formed. The values depend on $R$ to some extent.

In the case where $R = H$ (ethylene), the bond strength values are approximately 607 kJ/mol for the carbon–carbon double bond and 349 kJ/mol for the carbon–carbon single bond. In the system

$$
\begin{array}{ccc}
\text{H} & \text{H} & \text{H} \\
| & | & | \\
\sim\sim\text{C}-\text{C}-\text{C}\cdot & & \\
| & | & | \\
\text{H} & \text{H} & \text{H}
\end{array}
\qquad
\begin{array}{c}
\text{H}\!\!\diagdown\qquad\diagup\!\!\text{H}\\
\text{C}=\text{C}\\
\text{H}\!\!\diagup\qquad\diagdown\!\!\text{H}
\end{array}
$$

(growing polymer chain)    (monomer unit)<br>d.p. = $n$

the selective addition of $607/6.02 \times 10^{23}$ kJ would destroy the double bond in the monomer unit and leave the following:

$$
\begin{array}{ccc}
\text{H} & \text{H} & \text{H} \\
| & | & | \\
\sim\sim\text{C}-\text{C}-\text{C}\cdot & & \\
| & | & | \\
\text{H} & \text{H} & \text{H}
\end{array}
\qquad
\begin{array}{c}
\text{H}\!\!\diagdown\qquad\qquad\diagup\!\!\text{H}\\
\text{C:}\;\;\text{:C}\\
\text{H}\!\!\diagup\qquad\qquad\diagdown\!\!\text{H}
\end{array}
$$

If two of the electrons that were originally associated with the double bond were joined to form a single bond as shown:

$$
\begin{array}{ccc}
\text{H} & \text{H} & \text{H} \\
| & | & | \\
\sim\sim\text{C}-\text{C}-\text{C}\cdot & & \\
| & | & | \\
\text{H} & \text{H} & \text{H}
\end{array}
\qquad
\begin{array}{cc}
\text{H} & \text{H} \\
| & | \\
\cdot\text{C}-\text{C}\cdot \\
| & | \\
\text{H} & \text{H}
\end{array}
$$

this would release $349/6.02 \times 10^{23}$ kJ. Joining the electron on the end of the growing chain with one of the remaining two free electrons on the monomer unit would result in another single bond and the release of an additional $349/6.02 \times 10^{23}$ kJ:

$$
\begin{array}{ccccc}
\text{H} & \text{H} & \text{H} & \text{H} & \text{H} \\
| & | & | & | & | \\
\sim\sim\text{C}-\text{C}-\text{C}-\text{C}-\text{C}\cdot \\
| & | & | & | & | \\
\text{H} & \text{H} & \text{H} & \text{H} & \text{H}
\end{array}
$$

(growing polymer chain)<br>d.p. = $n + 1$

If the above process was repeated $6.02 \times 10^{23}$ times, the net energy output for adding one mole of monomer would be

607 kJ in

349 kJ out

349 kJ out

---

91 kJ out

The significance of this 91 kJ/mol can be seen by assuming that none of this heat escapes. It is known from physics that the increase in temperature, $\Delta T$, due to a heat input is given by*

$$\Delta T = \frac{q}{mc}$$

where $q$ is the heat, $m$ is the mass, and $c$ is the specific heat. Most organic systems have a specific heat (see Chapter 7) of approximately 2100 J/kg · K. If this is acceptable for both the polymer and the monomer in the example, then $\Delta T$ is given by

$$\Delta T = \frac{(91 \times 10^3 \text{ J/mol})}{(0.028 \text{ kg})(2.1 \times 10^3 \text{ J/kg · K})} = 1548 \text{ K}$$

It should be evident that this temperature is intolerable and, in fact, would destroy most polymers. During polymerization, therefore, one must be certain to remove this heat at a sufficient rate to limit the temperature rise. Although this example deals with polyethylene formation, the heat generated during the polymerization of other monomers can be calculated in a similar manner. Table 17.1 lists these values for several common polymers.

**TABLE 17.1  Heats of Polymerization of Some Common Polymers**

| Polymer | $\Delta H_P$ |
| --- | --- |
| Polypropylene | 85.8 kJ/mol |
| Polystyrene | 69.9 kJ/mol |
| Polyisoprene | 72.8 kJ/mol |
| Polyvinyl chloride | 95.8 kJ/mol |
| Polymethyl methacrylate | 56.5 kJ/mol |

(*Adapted from* [1].)

## POLYMERIZATION PROCESSES

As mentioned earlier in this chapter, the four major industrial methods for producing synthetic polymers are bulk, solution, emulsion, and suspension polymerization. In the latter three methods, polymerization occurs in a reactor in which a large ratio of surface area to volume is maintained for the monomer, and the monomer is in contact with either water or solvent (Figure 17.1). Heat control is relatively easy in these cases, since a large portion of the reactor volume contains inert material (the water or the solvent) that functions as a heat sink rather than a heat source. In addition, the effect of the low thermal conductivity

*This analysis ignores the temperature dependence of specific heat (Chapter 7).

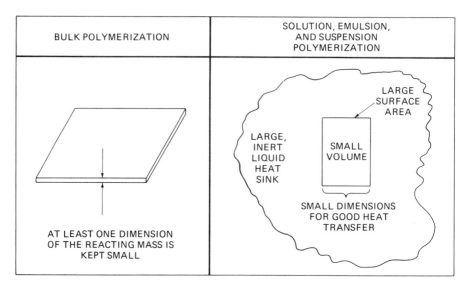

**Figure 17.1** Basic Principles of Heat Control.

of the polymerizing material is minimized because of the small distances involved. The differences among solution, emulsion, and suspension polymerization derive primarily from the methods used to produce the high surface-to-volume ratios. In bulk polymerization, on the other hand, no inert material is present to act as a heat sink. Heat control depends to a large extent, in this case, on keeping at least one dimension of the polymerizing mass small (Figure 17.1). The remainder of this chapter is a brief description of these four processes.

### Bulk Polymerization

In this process the monomer is polymerized without any significant amounts of other material present. Objects cast from epoxy or MMA are among the simplest examples of this process. On an industrial scale the formation of LDPE can be considered bulk polymerization. In this process ethylene gas (along with trace quantities of $O_2$) is pumped into a reactor at approximately 1500 atmospheres and $200°C$. Under these conditions the oxygen produces free radicals, which initiate the chain polymerization. As indicated in the previous section, this is a highly exothermic reaction. If this heat is not removed at a sufficient rate, explosive compounds (such as hydrogen and methane) could be created; and in this and other bulk systems, the relatively low conductivity of the monomer and the polymer accentuates this problem. In LDPE production, heat control is based on the use of narrow tubular reactors, which minimize two of the dimensions of the polymerizing mass. Although heat control is a major problem of bulk polymerization, this process does have some advantages over other processes. These advantages include:

1. The largest yield per reactor volume (since only monomer, polymer, and initiator are present)
2. The highest-purity product

The latter is particularly important in some optical and electronic applications where impurities cannot be tolerated.

### Solution Polymerization

In this process the monomer is dissolved in an inert solvent. In the production of polystyrene, for example, a typical formulation might include approximately 20% by weight of styrene monomer, approximately 80% by weight of benzene (solvent), and a small amount of initiator [2]. The purpose of the solvent is to absorb much of the heat generated during polymerization. If only styrene and initiator were present, the adiabatic rise in temperature would be approximately

$$\frac{Y\left(\frac{1 \text{ mol}}{0.104 \text{ kg}}\right)(69.9 \text{ kJ}/\text{mol})}{(2.1 \text{ kJ}/\text{kg} \cdot \text{K})(Y)} = 320 \text{ K}$$

where $Y$ is the total mass in kilograms. Replacing 80% of the monomer with solvent reduces by 80% the total amount of heat generated. If the heat capacity of the solvent is also $2100 \text{ J}/\text{kg} \cdot \text{K}$, then the adiabatic temperature rise is only [2]

$$\frac{(0.2 \, Y)\left(\frac{1 \text{ mol}}{0.104 \text{ kg}}\right)(69.9 \text{ kJ}/\text{mol})}{(2.1 \text{ kJ}/\text{mol} \cdot \text{K})(Y)} = 64 \text{ K}$$

Although the solution process has the advantage of better heat control than bulk polymerization, this is not obtained without paying a price. The disadvantages of the process are the following:

1. The solvents for most monomers are organic liquids and are therefore subject to handling problems such as toxicity, flammability, and pollution.
2. The solvents add a significant expense.
3. The yield per reactor volume is lower than in the bulk system, since the solvent takes up much of the space.
4. The resulting polymer contains trace quantities of solvent.

Nonetheless, solution polymerization plays an important role in polymer processing. Some systems (such as the polymerization of HDPE with hetero-geneous stereospecific catalysts) currently can be produced only by this method. Solution polymerization is also useful in systems where the final form for marketing is a polymer solution. This is the case, for example, with lacquers that are applied as polymer solutions and "dry" by solvent evaporation.

### Emulsion Polymerization

Given the choice, most people (and industries too) would prefer to work with water rather than with organic solvents. It would be ideal, therefore, if water could be used as the heat sink during polymerization rather than an organic solvent. The problem is that most monomers are hydrophobic and do not dissolve in water; but this can be circumvented by using emulsifiers (soaps, surfactants). These emulsifiers disperse hydrophobic monomer as droplets in water, just as household soaps do in the removal of oil or grease from clothing. The smallest of these soap-stabilized structures that contain monomer are called micelles, and polymerization takes place inside them. Since they are small and surrounded by water, the heat of polymerization is readily absorbed and controlled. Emulsion polymerization is a fairly complicated process; for details, see references [1] through [5].

The result of emulsion polymerization is a latex in which the hydrophobic polymer molecules are stabilized in water by the emulsifiers. In this form the polymer can be used as a water-based paint or an adhesive. The commercial white glues, for example, are primarily emulsions of polyvinyl acetate:

$$\left[\begin{array}{cc} H & H \\ | & | \\ -C & -C- \\ | & | \\ H & O \end{array}\right]_n$$
$$C=O$$
$$|$$
$$CH_3$$

They do not require stirring or shaking before use because a latex is thermodynamically stable.

In many applications, however, the solid polymer is the desired product, not the emulsion or latex. This solid could be obtained by spray-drying the latex to evaporate the water, but almost all of the soap would then remain as a contaminant. A preferred method of removing the water is to deactivate the soap by adding an acid. This causes the polymer particles to coagulate into a mass, which can then be further purified and dried. Even in this case one should recognize that traces of emulsifiers, which might limit the applications, will always exist.

### Suspension Polymerization

An alternative to emulsion polymerization that still gives one the advantages of having water as the heat sink is suspension polymerization. Unlike an emulsion, a suspension of monomer droplets in water is not stable and must constantly be stirred in order to prevent coalescence of the monomer or polymer droplets. A suspending agent is usually added to help segregate the droplets.

TABLE 17.2  Typical Polymerization Methods
for Common Polymers

| Polymer | Typical Polymerization Method |
|---|---|
| Low-density polyethylene | Bulk |
| High-density polyethylene | Solution |
| Polyvinyl chloride | Emulsion, suspension |
| Polystyrene | Bulk, suspension |
| Polymethyl methacrylate | Bulk, suspension |
| Polycarbonate | Bulk |
| Polypropylene | Solution |
| Nylon 6, other nylons | Bulk |
| Polyethylene terephthalate | Bulk |
| Polyisoprene | Solution |
| Polychloroprene | Emulsion |
| Styrene–butadiene rubber | Emulsion |
| Polytetrafluoroethylene | Suspension |
| Polyformaldehyde | Solution |

(*Adapted from* [4].)

Suspension polymerization has some advantages over emulsion polymerization. One is that the size of the polymeric product can be controlled by the amount of agitation. Hard stirring results in small droplets, less vigorous stirring in larger droplets. A second advantage is that the polymer is relatively easy to filter from the water, since the suspension is an unstable system. Disadvantages of this process compared to emulsion systems stem from the fact that the system must be stirred continually. More complete descriptions of this process and the others can be obtained by consulting the references at the end of this chapter.

It is important to remember that all polymers are not produced by all the processes just described. One is generally preferred for a particular system, depending on economic and production considerations, as well as the desired form of the polymer. Table 17.2 lists the more common methods of producing some polymers.

## PROBLEM SET

1. Using 607 kJ/mol as the strength of a carbon–carbon double bond and 349 kJ/mol as the strength of a carbon–carbon single bond, show that the heat of polymerization of ethylene is 91 kJ/mol.
2. What are the four industrial methods commonly used to produce polymer from monomer?
3. Which of the methods listed in Problem 2:
   (a) Produces the purest polymer?
   (b) Is the most difficult to control with respect to heat evolution?
   (c) Requires the use of large amounts of solvents?

(d) Produces a product containing surfactant residues that would prevent it from being used in some electrical or optical applications?

(e) Is used to produce latex paints?

(f) Provides the greatest possible yield per reactor volume?

4. Why do most monomers not dissolve readily in water?

5. What is a soap? How does it disperse hydrophobic materials (hydrocarbon oils, grease, some monomers, some polymers, etc.) in water? (Suggested reference: [2].)

6. How does water compare with typical organic solvents (acetone, benzene, methanol) as a heat sink?

7. Acrylic sheet is made by casting monomer between two smooth plates. It is an example of bulk polymerization. How is heat controlled in this process?

**Answer:**

Heat control in this process is accomplished by two methods. One is that a prepolymer (a solution of a polymer in its monomer) is generally used as the casting material. Therefore some of the polymer has already been formed prior to casting. The heat associated with that polymerization has been removed, and the polymer acts as an inert heat sink for the polymerization of the remaining monomer. The second heat-control mechanism is that one dimension of the sheet (its thickness) is relatively small. This means that (despite the low thermal conductivity of the polymerizing mass) the heat can readily escape without damaging the casting.

8. What basic principle do the polymerization methods (other than bulk) have in common?

9. How does the processing of chain and stepwise mechanism polymers differ with respect to when high-molecular-weight material can be drawn off?

10. Calculate the temperature rise associated with an adiabatic polymerization of one mole of each of the polymers listed in Table 17.1. Assume that $c = 2.1 \text{ kJ}/\text{kg} \cdot \text{K}$ in each case for both the monomer and the polymer.

11. If polymerization is exothermic, why is heat often added initially, as in the formation of acrylic sheet or in the production of plywood?

12. Shrinkage occurs during polymerization. This can be a problem, for example, when objects are cast from monomer, which is then polymerized in a mold. The degree of shrinkage can be estimated in these cases by comparing the densities of the monomer and the resulting polymer. The density of methyl methacrylate is approximately 0.9 g/cm³. Using the *Modern Plastics Encyclopedia* (see "Acrylic"), find the density of PMMA and calculate the shrinkage that occurs during polymerization. Neglect thermal expansion and contraction effects in your calculations.

13. How can the shrinkage of PMMA be minimized?

14. How does the shrinkage of epoxies during polymerization compare with that of MMA?

**Answer:**

The shrinkage of epoxies is a function of the type of curing agent used. Most exhibit relatively low shrinkages (approximately 4% for DGEBA/DTA in stoichiometric proportions) during polymerization, whereas the shrinkage for MMA is approximately

20%. This is an important consideration in precision castings. Fillers and prepolymers can be used to reduce shrinkage in specific systems.

15. Why does bond strength in vinyl monomers depend on $R$?
16. Distinguish among solutions, emulsions, and suspensions in terms of
    (a) Structure (definition)
    (b) Stability

## REFERENCES

[1] Rodriguez, F., *Principles of Polymer Systems*, McGraw-Hill, New York (1970).

[2] Rosen, S. L., *Fundamental Principles of Polymeric Materials*, Wiley, New York (1981).

[3] Flory, P. J., *Principles of Polymer Chemistry*, Cornell University Press, Ithaca, N.Y. (1953).

[4] Saunders, K. J., *Organic Polymer Chemistry*, Chapman and Hall, London (1973).

[5] Odian, G., *Principles of Polymerization*; McGraw-Hill, New York (1970).

# fabrication of parts and products

## fabrication processes

## chapter 18

Polymeric products are formed by a variety of fabrication techniques. Many of these techniques are described in detail in the *Modern Plastics Encyclopedia* [1], and in recent texts [2–9]. This chapter is intended as a brief introduction to these various methods and the types of products which they are typically used to produce.

## THERMOPLASTIC EXTRUSION

Figure 18.1 illustrates the major components of a typical extruder. In this apparatus, raw polymer (usually in pellet or powder form) is fed from a hopper to a rotating screw. The screw conveys the polymer through a series of zones in which the material is heated, compacted, and softened. The molten polymer is then forced by the screw through a die having a particular shape, and this shaped extrudate is then cooled to solidify it, either in air or in a bath. Products having uniform cross sections are frequently produced by extrusion. These include tubing, pipe, sheet, and insulating coatings for wire and cable. The raw materials for other fabrication processes are also sometimes produced by extrusion. Pellets, for example, can be made by first extruding a solid rod, and then chopping it to a specified length; and parisons (to be explained later) for blow-molded items such as polyethylene bottles are also usually extruded. A variation of extrusion is coextrusion, in which two or more materials are extruded through a common die. This technique is useful in producing various multilayer products such as packaging films.

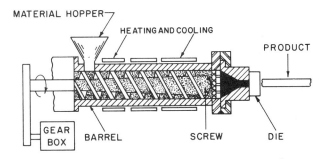

**Figure 18.1** Cross Section of an Extruder.

## BLOW MOLDING AND BLOWN FILM EXTRUSION

Blow molding and blown film extrusion are similar processes which utilize air to shape softened polymer tubes. Blow molding is used to produce hollow items such as milk and soda bottles, whereas blown film extrusion is used to produce relatively thin film and bags. In both cases a tube is first formed in a shape that approximates the article to be made. In the case of bottlelike objects, the tube (called a parison) is either extruded or injection-molded and is then clamped in a split mold with a blow pin inside. Air inflates the polymer until it conforms to the shape of the mold. The polymer cools on contact with the cold mold walls, the mold is then opened, and the part is ejected (Figure 18.2). In blown film extrusion no mold is required (Figure 18.3). Instead, a continuous tube is extruded, clamped between rollers at one end, and blown with compressed air (much like an oblong balloon) to form a large, hollow tube having a very thin wall (5 mils, for example). This tube can be continuously flattened and placed on a roll for bag production, or it can be slit, opened, and used for film.

**Figure 18.2** Blow Molding.

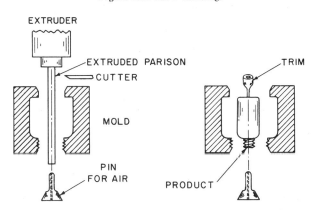

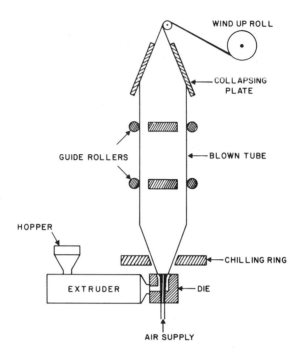

**Figure 18.3** Blown Film Extrusion. (*Adapted from* [1].)

## THERMOPLASTIC INJECTION MOLDING

Injection molding involves shooting or injecting a charge of molten polymer into a cold mold. Once the material has solidified in the mold, it is removed and the cycle is repeated. A crucial part of the injection-molding process is the preparation of the molten charge, which is typically done with either a plunger-type injector (Figure 18.4) or a reciprocating screw injector (Figure 18.5). In a plunger-type injector, a preplasticating screw extruder softens the polymer and

**Figure 18.4** Plunger-Type Injection Molder with Preplasticating Screw. (*Adapted from* [1].)

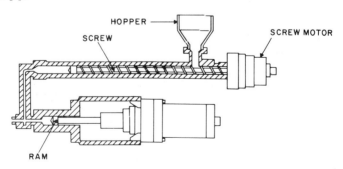

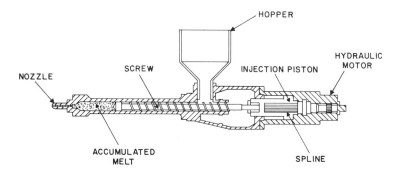

**Figure 18.5** Reciprocating Screw Injection Molder. (*Adapted from* [1].)

conveys it to a chamber that directly precedes a hydraulic ram. At the proper time the ram moves forward and forces the accumulated molten polymer into the mold. The ram then retracts, permitting the screw to refill the chamber. A reciprocating screw injection molder also has a screw that softens and conveys the polymer molding compound; but in this case the screw is also the ram. At the proper time the screw moves forward, injecting the accumulated melt into the mold.

Injection molding is adaptable to very high production rates and rather complex shapes. The cycle times are relatively short (on the order of seconds), and multicavity molds are commonly used for producing small parts. The size of the charge can vary from about 2 oz to 800 oz, depending on the size of the injection molder. Many common products are frequently produced by injection molding, including containers such as cups, dishpans, buckets, flower pots, and garbage cans; machine parts such as gears, housings, and bearings; and personal products such as combs, pens, and disposable razors.

## ROTATIONAL MOLDING

In rotational molding, a powder or a liquid plastisol is placed into a cold hollow mold. As this mold is heated, it is rotated simultaneously on two perpendicular axes. This causes the polymer to soften or melt and coat the inside of the mold and form a hollow object. While the mold is still spinning, it is then cooled; and when the polymer has solidified, the object is removed from the mold. A typical rotational molder has three stations: one for loading and unloading, one for heating, and one for cooling (Figure 18.6).

An important feature of rotational molding is that several different items can be produced from different materials at the same time on the same machine. For example, one arm of the molder could support a single mold for producing a large polyethylene tank, while the second arm supports several smaller molds for producing polyethylene pipe fittings and a third supports several dozen very small molds for producing a toy from a plastisol. This versatility makes

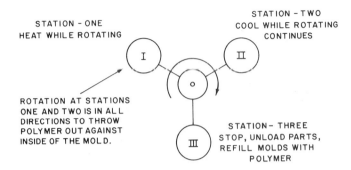

STATION - ONE
HEAT WHILE ROTATING

STATION - TWO
COOL WHILE ROTATING
CONTINUES

ROTATION AT STATIONS
ONE AND TWO IS IN ALL
DIRECTIONS TO THROW
POLYMER OUT AGAINST
INSIDE OF THE MOLD.

STATION- THREE
STOP, UNLOAD PARTS,
REFILL MOLDS WITH
POLYMER

**Figure 18.6** Rotational Molding Operation.

rotational molding useful not only for low-volume production of large items but also for high-volume production of relatively small parts. The most common raw materials currently used in rotational molding include PVC (as a plastisol), polyethylene (including crosslinkable grades), polycarbonate, nylon, and vinyl acetate.

## CALENDERING

Calendering of thermoplastics is somewhat similar to the rolling of metal sheet. The product is formed by a series of heated, polished rollers that flatten, extend, and draw the molding compound (Figure 18.7). A calender is relatively large and expensive, but its production rate is correspondingly high. Embossed rollers can be utilized to produce textured surfaces on the films. Some common products produced by calendering include shower curtains, tablecloths, upholstery, floor tile, and woodgrain laminating film. In addition to PVC, ABS, ABS/PVC blends, and polyethylene are frequently calendered.

POLYMER

**Figure 18.7** Calendering Process.

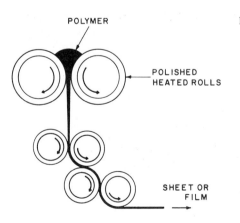

POLISHED
HEATED ROLLS

SHEET OR
FILM

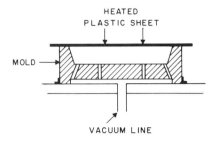

MOLD

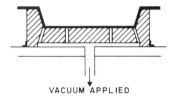

**Figure 18.8** Sheet Forming Using Vacuum.
(*Adapted from* [1].)

FORMED PART

## SHEET FORMING (Thermoforming)

This process is used to produce a three-dimensional object from a flat polymer sheet. The thermoplastic sheet is first heated to its softening point, then shaped to a form, and finally cooled to retain the shape (Figure 18.8). Shaping of the sheet is done by mechanical force, air pressure, vacuum, or some combination of these. Many variations of the basic process shown in Figure 18.8 have been developed, such as matched die forming, drape forming, and plug assist forming. Polystyrene is the most frequently used thermoforming material, but other thermoformed polymers include ABS, acrylic, polycarbonate, PVC, high-density polyethylene, and cellulosics. Many common products are produced by thermoforming methods, including egg cartons (from expanded polystyrene foam), refrigerator liners, disposable dinnerware, automotive panels, and skylights.

## COMPRESSION MOLDING

In compression molding a hydraulic or pneumatic press with heated platens is utilized to shape and fully polymerize a molding compound in a hot mold. This is a cyclical process in which a molding compound is placed in a heated mold; the

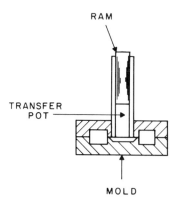

MOLD                    **Figure 18.9** Transfer Mold.

press is then activated to close the mold and shape the charge as it polymerizes; and, finally, the press is opened to remove the finished part. The molding compounds are typically thermosets such as filled or reinforced phenol formaldehydes, melamines, or thermosetting polyesters. The use of thermoplastics is quite limited, largely because of the inefficiencies inherent in repeated heating and cooling of the mold and the press platens. Some products commonly produced by compression molding include building panels, dinnerware, machine housings, and electrical components.

A variation of compression molding is transfer molding. In this process the preheated molding composition is first placed in a separate chamber or pot (often using a plasticating screw), rather than directly into the mold. As the press closes, the molding composition is injected into the mold (Figure 18.9). This method has several advantages over conventional compression molding, including lower mold wear, since no solid material is forced through the mold.

## THERMOSET INJECTION MOLDING
## AND REACTION INJECTION MOLDING

Traditionally, injection molding has been limited to thermoplastics, but in recent years the injection molding of thermosets has become more common. As in thermoplastic injection molding, this involves injecting a molten charge into a mold. Thermoset injection molding is fundamentally different from thermoplastic injection molding, however, because of the reactive nature of the thermosetting charge. The equipment is somewhat different. For example, preplasticating screws are designed so as to minimize problems with polymerization in the molder, and the molder itself is designed for rapid cleanout. However, the major difference is in the mold, which must be heated (rather than cooled) in order to polymerize the charge.

A variation of thermoset injection molding is reaction injection molding (RIM). In this case, liquid reactants (an isocyanate and a polyol, for example) are

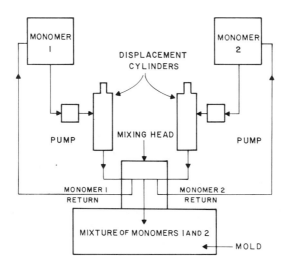

**Figure 18.10** Reaction Injection Molding System. (*Adapted from* [1].)

used instead of granular material. These two liquids are injected separately into a mixing head, and from this mixing head into a warm mold (Figure 18.10). Because the materials are very reactive liquids, the molding times and mold pressures are relatively low. Most of the current applications for this process are in the transportation industry and include air scoops, fender extensions, and bumpers.

## PULTRUSION

Pultrusion is the thermosetting equivalent of thermoplastic extrusion, in that both can produce products of constant cross section and infinite length. They are fundamentally different, however, in how this product is achieved. In a pultrusion unit, continuous fibers (usually glass) are drawn through a bath of thermosetting monomer. These resin-coated fibers are then drawn through an elongated, heated, steel die, which shapes the mixture and polymerizes the resin. Polyesters are the most common polymers utilized in pultrusion, although others such as one-component epoxies are used as well. Products produced by this method include electrical equipment such as safety ladders, and sporting goods equipment such as fishing rods and CB antennas.

## FILAMENT WINDING

In principle, filament winding is a relatively simple process in which resin-soaked continuous fibers are wound around a mandrel of a given shape. The resin is then polymerized and the mandrel removed (Figure 18.11). The machinery can be

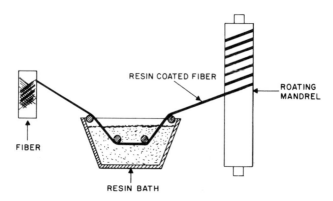

**Figure 18.11** Filament-Winding System. (*Adapted from* [1].)

very sophisticated, however. Many filament winders are preprogrammed with instructions for rather complex patterns engineered to resist anticipated stresses.

Glass fiber is the most common reinforcing agent used in filament winding, although graphite and aramid fibers are also used. The resins are typically epoxies, thermosetting polyesters, silicones, and phenolics. Products produced by filament winding include tanks, pipe, rocket casings, and various airplane components.

## HAND LAYUP

Hand layup is the sequential buildup of layers of fiber-reinforced polymer over an open mold or form. The process usually includes a coating of unreinforced resin (termed a gel coat), followed by layers of reinforced resin. The reinforcement is usually glass fibers, and the resin is typically either a room-temperature curing thermosetting polyester or epoxy. After the resin has polymerized, the structure is removed from the form or mold. Hand layup is a relatively low-volume process, but one particularly well suited to large items such as boat hulls. Modifications of this process include sprayup, in which chopped fiber reinforcement is simultaneously sprayed on the form or in the mold with the resin.

## CASTING METHODS

Casting involves pouring a liquid into a mold and allowing it to harden. The liquid can be either a liquid monomer, a solution of polymer in its own monomer (prepolymer), a polymer melt, or a solution of polymer in an inert solvent. In each case the solidification process is different. These are compared in Table 18.1. Casting is usually used for low-volume production, although some exceptions (such as cast acrylic sheet) exist.

**TABLE 18.1 Casting Methods**

| Method | Starting Material | Solidification Method | Examples | Comments |
|---|---|---|---|---|
| Thermoset casting | Monomer ($f > 2$) | Room- or elevated-temperature polymerization | Epoxy castings (e.g., encapsulation of electronic components) | Heat of polymerization must be controlled |
| Thermoplastic casting | Monomer or prepolymer | Polymerization at elevated temperature | Acrylic sheet castings | |
| | Melt | Cool below $T_m$ and/or $T_g$ | Unoriented polypropylene film, nylon parts | |
| Solvent casting | Solution of polymer in an inert solvent | Evaporate solvent | PVC film | |

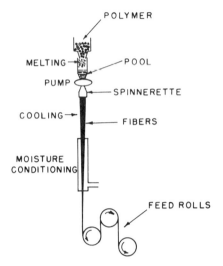

**Figure 18.12** Melt Spinning of Fibers.

## FIBER PRODUCTION

Fibers for use in textile and other applications are usually produced either by melt spinning, dry spinning, or wet spinning. All are extrusion processes, but the particular technique utilized depends on the polymer involved. Nylons, for example, are typically melt-spun. In this process the molten polymer is extruded through a spinnerette (a plate containing many small holes). After the extrudate from these holes cools somewhat, it is drawn by a series of rolls to enhance and orient the crystallinity (see Chapter 5). The resulting fibers are then spun to make thread for weaving and other operations (Figure 18.12). In dry spinning, a polymer solution (rather than a polymer melt) is extruded through the spinnerette holes. Air currents evaporate the solvent, leaving the polymer in a fiber form which is then drawn. Polyacrylonitrile fiber is typically produced by this process. Wet spinning is similar to dry spinning but differs in that the extrudate is drawn through a liquid, which causes the polymer to precipitate from solution. Rayon fiber, for example, is typically wet-spun.

## PROBLEM SET

1. Characterize the processes described in this chapter according to:
   (a) Whether they are appropriate to high or low production volumes
   (b) Whether they are continuous or intermittent processes
   (c) Whether they are used primarily to fabricate thermoplastics or thermosets
2. Characterize the processes described in this chapter according to their limitations on part geometry (primarily for hollow objects, for products of uniform cross section, etc.).

3. Suggest a process for producing the following. Support your choice by contrasting it with other possible alternatives.
   (a) Polyethylene milk bottles
   (b) Nylon film
   (c) Polystyrene egg cartons
   (d) Seamless polyethylene gas tank for a tractor
   (e) Copper wire with a polyvinyl chloride insulation
   (f) Distributor caps containing molded-in contacts
   (g) A small number of nylon gears
   (h) A large number of nylon gears
   (i) Polyvinyl chloride garden hose
   (j) Polyethylene ice cube trays

4. Explain how it is possible to extrude a parison of variable cross section for blow molding. (Suggested reference: [1].)

5. What is the major disadvantage of compression molding thermoplastics? What are some typical thermosetting items that are compression-molded?

6. Compare and contrast the critical features of pultrusion and thermoplastic extrusion. See the *Modern Plastics Encyclopedia* [1] for detailed descriptions of these processes.

7. Distinguish between blown film extrusion and blow molding.

8. Some thermosetting compositions are solids (e.g., phenolic and melamine molding powders), whereas others are liquids (e.g., isocyanates and polyols) before they are molded into parts. Explain how this is possible.

9. Describe (in words and with a diagram) the thermoplastic extrusion process. Cite three common examples of products made by this method.

10. Repeat Problem 9 for the following fabrication processes:
    (a) Thermoplastic injection molding
    (b) Calendering
    (c) Rotational molding
    (d) Filament winding
    (e) Hand layup
    (f) Reaction injection molding

11. A related fabrication technique is the welding of polymers. What types of welding are common?

**Answer:** Common welding techniques for polymers include the following:
    (a) Ultrasonic welding, which utilizes mechanical energy at frequencies above human detection levels. Heat is generated by friction due to the rubbing of the polymer surfaces, and this fuses the two materials together.
    (b) Hot-gas welding, which simulates that of metal technology but does not utilize an open flame. Hot gas melts the edges of the surfaces to be joined, and a filler rod of the same composition is used to fill the gap. Because of the high viscosities of some polymer melts, pressure is often applied with a roller to aid in the fusion process.
    (c) Hot-wire welding, in which wires embedded in a joint are heated electrically to fuse the edges of the joint. The wire remains in the joint after cooling.
    (d) Friction or spin welding, in which mated parts are rotated and the contact surfaces melt and fuse together as a result of the frictional heat.
    (g) Solvent welding (see Chapter 11).

12. In Chapter 5 it was noted that the extrusion of a molten polymer can produce orientation in a product. This results in an anisotropic material (the properties vary with direction). Contrast the orientation that can occur in blow-extruded film with that which can occur in film extruded from a slot die.

13. Consider the production of polyethylene gas tanks by rotational molding. How would the following properties be important in this process?
    (a) Heat capacity
    (b) Thermal expansion
    (c) Conductivity
    (d) $T_m$
    (e) Melt viscosity
    How would the following properties be important in the final product?
    (a) Oxidative resistance
    (b) Modulus
    (c) Tensile strength
    (d) Impact strength
    (e) Solvent resistance
    (f) Electrical conductivity

## REFERENCES

[1] *Modern Plastics Encyclopedia*, vol. 57, no. 10A, McGraw-Hill, New York (1980–1981).

[2] Beck, R. D., *Plastic Product Design*, 2nd ed., Van Nostrand Reinhold, New York (1980).

[3] Throne, J. L., *Plastics Process Engineering*, Marcel Dekker, New York (1979).

[4] Tadmor, Z., and C. C. Gogos, *Principles of Polymer Processing*, Wiley-Interscience, New York (1979).

[5] Fenner, R. T., *Principles of Polymer Processing*, Chemical Publ. Co., New York (1980).

[6] Becker, W. E., *Reaction Injection Molding*, Van Nostrand Reinhold, New York (1979).

[7] Dym, J. B., *Injection Molds and Molding*, Van Nostrand Reinhold, New York (1979).

[8] Rubin, I. I., *Injection Molding—Theory and Practice*, Wiley-Interscience, New York (1973).

[9] DuBois, J. H., and W. I. Pribble, *Plastics Mold Engineering Handbook*, 3rd ed., Van Nostrand Reinhold, New York (1978).

Relatively few polymers today are used in their pure form; instead, most contain or are combined with other materials as well. Some of the reasons for doing this are the following:

1. To improve mechanical properties such as modulus, strength, hardness, abrasion resistance, and toughness
2. To prevent degradation (both during fabrication and in service)
3. To change the thermal properties, such as the expansion coefficient and the conductivity
4. To reduce materials costs
5. To improve the processability

Many of the materials that are mixed with polymers are called additives. Some of these have been mentioned in previous chapters. Additives are solids or liquids used primarily as colorants, UV absorbers, plasticizers, flame retardants, thermal stabilizers, lubricants, and antistatic agents. The resulting mixture can be either homogeneous or heterogeneous, depending on the solubility of the additive in the polymer. In any case, additives are an important part of the polymer industry. For example, phthalate-based plasticizer production* in the United States exceeds the combined production of epoxy and acrylic polymers.

Besides additives, a variety of particulate and fibrous materials are also frequently mixed with polymers. The result is typically a heterogeneous mixture;

*These are used primarily in the production of plasticized vinyl products.

167

that is, these materials do not dissolve in the polymer, and the result is a multiphase system. Fibers are materials with lengths many times their thicknesses and widths. Some that are frequently mixed with polymers [1] are

1. Glass
2. Carbon and graphite
3. Cellulosics such as alpha cellulose
4. Synthetic polymers such as nylon
5. Metals
6. Boron

Fibers are frequently added to polymers to increase strength, and under such circumstances they are called reinforcing agents. It is important to note, however, that adding fibers to a polymer does not always increase strength.

Particulates are a much more diverse group of materials than fibers, and they represent a wide range of geometries. Some are needlelike (tapered on both ends); some are saucer- or disk-shaped; some are almost perfect spheres; and others have irregular shapes. The more common particulates [1] include

1. Calcium carbonate
2. Wood flour
3. Carbon blacks
4. Glass spheres, flakes, and granules
5. Metallic powders and flakes
6. Silicate minerals such as clay, talc, and mica
7. Silica minerals such as quartz, diatomaceous earth, and novaculite
8. Metallic oxides such as alumina
9. Other synthetic polymers

Particulates are usually added to polymers for reasons other than to increase strength, and they are termed fillers in those situations. It is important to note, however, that some particulates are also used as reinforcing agents. For example, carbon black is a common reinforcing agent for elastomers.

When additives, fillers, or reinforcing agents are mixed with a polymer, they extend the useful range of properties of that polymer. At the same time, of course, they make the analysis of these materials more complex. The polymer itself has many variables that affect its properties, as noted in previous chapters. Since the mixture includes the polymer, all of these variables are still influential. In addition, however, others exist. In the case of a fibrous material embedded in a polymer matrix, for example, these extra variables include

1. The properties of the fiber (density, hardness, strength, etc.)
2. The size and shape of the fiber
3. The amount of fiber added
4. The alignment of the fiber
5. The distribution of the fiber
6. The nature of the interface between the polymer and the fiber

Although an analysis of the above variables is beyond the scope of this text, a relatively simple example will illustrate some of the versatility (and complexity) of these systems.

Consider the case of parallel fibers embedded in a polymer matrix, as shown in Figure 19.1. The direction of alignment is also the direction of the applied stress. The modulus of this composite in the fiber direction can be estimated by the rule of mixtures:

$$E_c = E_m v_m + E_f v_f$$

where the subscripts c, m, and f refer to the composite, matrix, and fiber, respectively, and $v_m$ and $v_f$ are the volume fractions of the matrix and the fiber. For the case in which there is no fiber (i.e., pure polymer) this equation reduces to

$$E_c = E_m(1) + E_f(0)$$

or

$$E_c = E_m$$

For the case in which there is no polymer, the equation reduces to

$$E_c = E_m(0) + E_f(1)$$

or

$$E_c = E_f$$

A specific example of this type of system is an epoxy matrix (amine-cured) containing long carbon fibers. Tables 19.1 and 19.2 list some typical properties for these two materials as they exist separately. The data in these two tables (along with the rule of mixtures) can be used to calculate the modulus of this

**Figure 19.1** Unidirectional Fiber-Reinforced Composite.

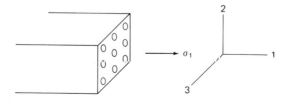

**TABLE 19.1 Properties of a Typical
Amine-Cured Epoxy Resin**

| | |
|---|---|
| Resin: | Epon 828* |
| Curing agent: | $m$-phenylenediamine |
| | (stoichiometric) |
| Density: | 1.18 g/cm$^3$ |
| Tensile strength: | 85.5 MPa |
| Tensile modulus: | 3.31 × 10$^3$ MPa |
| Tensile elongation to break: | 5.1% |

*Epon 828 is a Shell Chemical Company product.

(*Adapted from* [2].)

**TABLE 19.2 Properties of Carbon Fibers Produced from PAN (polyacrylonitrile)**

| Material | Density (g/cm$^3$) | Tensile Strength (MPa) | Young's Modulus (10$^3$ MPa) |
|---|---|---|---|
| High-modulus carbon fiber | 1.9 | 2100 | 390 |
| High-strength carbon fiber | 1.9 | 2500 | 240 |

(*Adapted from* [3].)

theoretical composite for any volume fraction of fiber. If the volume fraction was 0.6, for example, the modulus of the composite in the fiber direction would be

$$E_{c_l} = E_m v_m + E_f v_f$$

$$= (3.31 \times 10^3)(0.4) + (390 \times 10^3)(0.6)$$

$$= 235 \times 10^3 \text{ MPa}$$

While this value is approximately equal to that of steel, the density of the composite is much lower. The efficiency of this system can be compared to that of steel using the modulus-to-density ratio:

$$\frac{\text{Modulus}}{\text{Density}} = \frac{235 \times 10^3 \text{ MPa}}{(0.4)(1.18) \text{ g/cm}^3 + (0.6)(1.9) \text{ g/cm}^3} = \frac{235 \times 10^3 \text{ MPa}}{1.61 \text{ g/cm}^3}$$

$$= 14.6 \times 10^4 \text{ MPa/(g/cm}^3)$$

This is much better than steel, which has a ratio of

$$\frac{207 \times 10^3 \text{ MPa}}{7.8 \text{ g/cm}^3} = 2.65 \times 10^4 \text{ MPa/(g/cm}^3)$$

On a weight basis, therefore, the composite may be a better choice in applications where weight is critical. This is a major reason for the interest in these materials shown by the automobile, aerospace, and defense industries.

One more point should be made with regard to this example. The advantages gained in the direction of the fibers are offset to a large extent by the effects in the directions perpendicular to the fibers. If all the fibers are aligned as in Figure 19.1, the stiffness of this system in the other two principal directions is given by

$$E_{c_2} = E_{c_3} = \frac{E_f E_m}{E_f v_m + E_m v_f}$$

$$= 8.2 \times 10^3 \text{ MPa}$$

A unidirectional composite is much like a piece of wood: very stiff along the grain, but relatively compliant in directions perpendicular to the grain. This effect can be reduced by making sandwiches or laminates with layers having different fiber directions, similar to the way in which plywood is constructed.

Unidirectional fiber-reinforced polymers, as well as those containing particulates and fibers of various orientations, belong to a broad class of materials known as composites. In general terms, a composite is a heterogeneous mixture of two or more materials. This is in contrast to a homogeneous mixture, in which the materials do not exist in discrete phases. For example, an externally plasticized polymer is a homogeneous mixture or solution, but the same polymer containing an insoluble material is a composite. Composites can assume many forms, and the filled and reinforced polymers discussed in this chapter represent only one of these forms. Laminates (layers of different materials) are another form of composite, as are impregnated materials (in which one material is deposited in the pore structure of another material). Information concerning this diverse group of materials is available in the references listed at the end of this chapter [3–7].

## PROBLEM SET

1. Used the *Modern Plastics Encyclopedia* [8], find a supplier for the following additives, fillers, and reinforcing agents:
   (a) Asbestos staple fiber
   (b) Ground limestone filler
   (c) Antistatic agents
   (d) Glass spheres
2. What are ten reasons (other than mechanical) for adding fibers and particulates to polymers?
3. Let $E_m = 3.5 \times 10^3$ MPa and $E_f = 70 \times 10^3$ MPa for a unidirectional, continuous fiber composite. On one graph, plot $E_c / E_m$ versus fiber volume fraction for directions parallel and perpendicular to the fibers.
4. What does the graph in Problem 3 tell you about utilizing fibers efficiently?
5. Given a unidirectional, continuous steel fiber epoxy matrix composite as shown with the following parameters, calculate $E_{c_1}$, $E_{c_2}$, and $E_{c_3}$.

*Fiber:*      $E_f = 207 \times 10^3$ MPa

              $\rho = 7.80$ g/cm$^3$

*Composite:* $v_f = 0.50$

*Matrix:*    $E_m = 1.38 \times 10^3$ MPa

              $\rho = 1.18$ g/cm$^3$

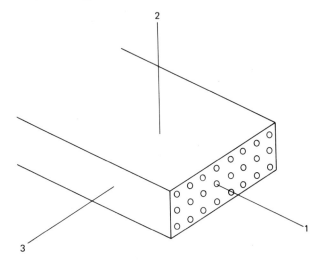

6. If the composite material described in Problem 5 is manufactured as follows, calculate $E_{c_1}$ and $E_{c_2}$.

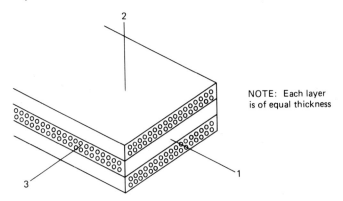

NOTE: Each layer
is of equal thickness

7. Assume that a beam of rectangular cross section is subjected to a bending moment. If the width is constant at 6 in, how deep would the beam have to be in order to obtain an equivalent stiffness ($EI = 3.24 \times 10^9$ lb-in$^2$) for the following materials?

*Steel:*

$$E = 207 \times 10^3 \text{ MPa}; \ \rho = 7.80 \text{ g/cm}^3$$

*Aluminum*:

$$E = 70 \times 10^3 \text{ MPa}; \ \rho = 2.71 \text{ g/cm}^3$$

*Eastern hemlock No. 1 at 15% M.C.*:

$$E = 9.0 \times 10^3 \text{ MPa}; \ \rho = 0.45 \text{ g/cm}^3$$

*Epoxy*:

$$E = 3.5 \times 10^3 \text{ MPa}; \ \rho = 1.18 \text{ g/cm}^3$$

*Epoxy/S glass composite*:

$$v_f = 0.35; \ E = 70 \times 10^3 \text{ MPa}; \ \rho = 2.15 \text{ g/cm}^3$$

**8.** Compare the weights (per unit of beam length) for each of the alternatives in Problem 7. How do these relate to the modulus-to-density ratios of the various materials?

**9.** The amount of fiber or particulate that can be added to a polymer matrix is limited by geometrical considerations. The following are the densest packing arrangements for fibers and spheres of uniform size. Calculate the volume fraction of fibers or spheres in each case.

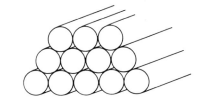

MAXIMUM PACKING FOR UNIFORM FIBERS

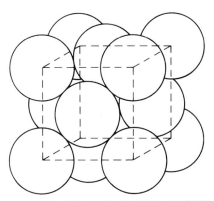

MAXIMUM PACKING FOR UNIFORM SPHERES
(cube face diagonal = 4 × sphere radius)

**10.** The arrangements shown in Problem 9 yield the highest packing fractions. What can be done to increase this value?

11. Adding fibers to polymers can affect more than just modulus or strength. Using the Plastics Properties Chart from the *Modern Plastics Encyclopedia* [8], compare unmodified type 6 nylon with 30–35% glass-reinforced type 6 nylon with regard to:
    (a) Density
    (b) Tensile strength
    (c) Elongation to break
    (d) Impact strength
    (e) Thermal expansion

12. What are some commercial products that utilize carbon fibers embedded in an epoxy matrix?

13. Although polystyrene is one of the highest-volume thermoplastics in use today, its impact strength is too low for many applications. A more impact-resistant form of polystyrene is HIPS (high-impact polystyrene). HIPS is a mixture or blend of a rubbery particulate (usually polybutadiene rubber) and polystyrene. The rubbery phase is dispersed in the polystyrene matrix, thereby making a composite that is less crack-sensitive and more impact-resistant than polystyrene. How would you expect the following other properties to change because of the presence of the rubbery phase?
    (a) Modulus
    (b) Clarity
    (c) Elongation to break
    (d) Strength
    Check your answers, using the *Modern Plastics Encyclopedia* [8].

## REFERENCES

[1] Seymour, R. B., *Modern Plastics Technology*, Reston Publishing Co., Reston, Va. (1975).

[2] *Epon Resins for Casting*, Shell Chemical Co. (1967).

[3] Argarwal, B. D., and L. J. Broutman, *Analysis and Performance of Fiber Composites*, Wiley, New York (1980).

[4] Manson, J. A., and L. H. Sperling, *Polymer Blends and Composites*, Plenum, New York (1976).

[5] Ashton, J. E., J. C. Halpin, and P. H. Petit, *Primer on Composite Materials*, Technomic, Stamford (1969).

[6] Jayne, B. A., *Theory and Design of Wood Fiber Composites*, Syracuse University Press, Syracuse, N.Y. (1972).

[7] Nielsen, L. E., *Mechanical Properties of Polymers and Composites*, vols. 1 and 2, Dekker, New York (1974).

[8] *Modern Plastics Encyclopedia*, vol. 57, no. 10A, McGraw-Hill, New York (1980–1981).

# design considerations

## chapter 20

## INTRODUCTION

The design of parts and finished products involving synthetic polymers is in many respects similar to that used for other materials. In each case, one is attempting to put the best material(s) in the best configuration, for the least cost. Elements of good design include not only consideration of the economics and performance of the material in the finished part but also consideration of how the part may best be produced. The finest materials in the world are of little use if parts and products cannot be economically fabricated from them.

Examples of good (and poor) design exist in all areas to which polymers are applied. However, some of the most innovative uses in recent years have occurred in the automotive industry. Because of the emphasis on fuel economy and the concomitant reductions in vehicle weight, more attention is being placed on the replacement of iron and steel with polymers and light-weight metals. Table 20.1 illustrates how the use of some of these materials is expected to change in the next few years.

The replacement of metallic components with polymeric ones in the automotive or any other area presents a number of problems as well as opportunities. One of the most critical problems in general is the lack of a large data or performance base for these polymeric materials and polymer-composite materials. For most of the metals there are decades of experience on which to rely in design. Most of the polymers, on the other hand, are comparatively new, so that extensive information about their long-term in-service behavior is not available.

**TABLE 20.1  Estimated Change in Material Usage from 1978 to 1987 for an Average Automobile**

| Material | Amount (Pounds) | | Changes (%) |
|---|---|---|---|
|  | 1978 | 1987 |  |
| Polymers (excluding rubber) | 189 | 255 | +35 |
| Aluminum | 121 | 180 | +49 |
| Steel | 2066 | 1687 | −18 |
| Iron | 625 | 283 | −55 |
| Rubber | 99 | 81 | −18 |
| Glass | 95 | 79 | −17 |
| Lead | 22 | 22 | 0 |
| Zinc | 18 | 9 | −50 |
| Other* | 264 | 210 | −20 |
|  | 3499 | 2806 | −20 |

*Copper, fabric, paper, fluids, lubricants, etc.*
(*Adapted from* [1].)

## BASIC PRINCIPLES

Whether one is designing for an automotive application or some other use, several principles hold. The first is that one should resist the temptation simply to substitute on a one-to-one basis. When the problem involves redesigning a part and utilizing a new material (e.g., substituting a polymer for a metal), one should

**Figure 20.1** Steel Tackle Box.

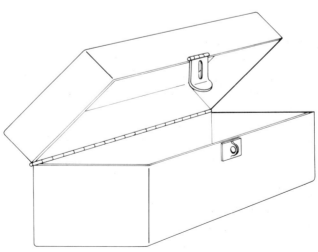

approach this from the standpoint of maximizing the benefit that can be derived from the properties of the new material. This can sometimes involve drastic changes in the original product. Consider the following rather elementary example. The small, trayless steel tackle box illustrated in Figure 20.1 consists of approximately a dozen parts, including a handle, hinges for the handle, a top, a bottom, a main hinge, and a latch mechanism (Figure 20.2). In redesigning this for a polymer equivalent, one would never try to duplicate each part and join them together. The parts as they exist in Figure 20.2 reflect the nature of the steel, and the economics of producing that item in steel. By selecting a suitable polymer (polypropylene, which in the oriented state can form a continuous, fatigue-resistant hinge), it is possible to reduce the tackle box to two parts: a handle and the rest of the tackle box (Figure 20.3). This example is a simple application of the combined function concept. Numerous everyday examples of this principle exist [2, 3].

Another basic principle of good design is the use of an extensive list of the

**Figure 20.2** Steel Tackle Box (Exploded View).

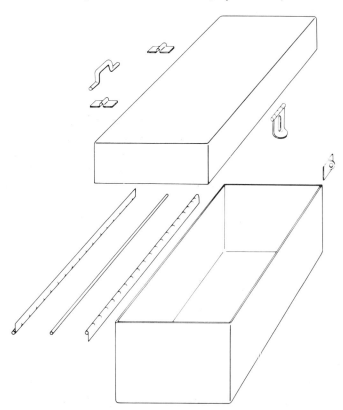

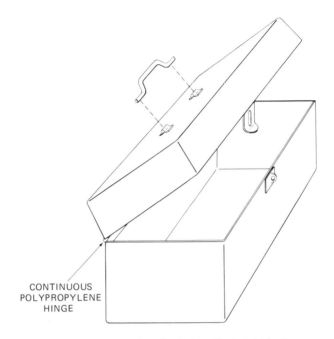

CONTINUOUS
POLYPROPYLENE
HINGE

**Figure 20.3** Polypropylene Equivalent (Exploded View).

properties required for a particular application. The importance of this in designing with polymers cannot be overemphasized, since polymers are relatively new and very diverse materials. Although the contents of these lists are obviously a function of the application, Table 20.2 illustrates some of the properties that can frequently appear.

It is in the development of such a list that the importance of what has been discussed in Chapters 1 through 19 becomes most evident. Polymers are very different from other materials, and they also differ widely among themselves. In particular, their viscoelastic tendencies (Chapter 13), their relatively high coefficients of thermal expansion (Chapter 7), their combustibility (Chapter 10), their susceptibility to sunlight and other environmental factors (Chapter 9), their limited range of use temperatures (Chapter 8), and their resistance and susceptibility to various solvents (Chapter 11) may all require more detailed attention in the design process than has been shown in the past with other materials. The problems, of course, become compounded when composite systems are considered (Chapter 19). Thus it is important to write down the requirements. This will not provide a final choice of a material, but it will narrow the field. Final material decisions should always involve testing of prototypes.

Proper and efficient utilization of polymers in design requires an under-

TABLE 20.2  Some Items That Typically Appear on a
List of Product Requirements

| Mechanical | Appearance |
|---|---|
| 1. Stress types | 1. Shape restrictions |
| 2. Stress magnitudes | 2. Color |
| 3. Loading rate | 3. Finish |
| 4. Allowable deformation | |
| **Environmental** | **Economics** |
| 1. Temperature requirements | 1. Present cost of item being |
| 2. Solvent exposure | replaced |
| 3. Acid or base exposure | 2. Expected volume |
| 4. Ozone or other atmospheric | 3. Equipment restrictions |
| contaminants | 4. Cost ceiling |
| **Special Requirements** | |
| 1. Electrical (codes, properties) | |
| 2. Sterilizable or steam-cleanable | |
| 3. Permeability | |
| 4. Flammability | |
| 5. Ultraviolet or other radiation resistance | |

standing of their properties and behavior, and the purpose of this text has been to act as an introduction to this topic. More information with regard to the use of polymers can be obtained from a number of sources, including those listed in previous chapters. In particular, the following periodicals often contain examples of design problems:

*Modern Plastics*
*Plastics Design Forum*
*Plastics Engineering*

Several textbooks emphasizing design are also available [4–6].

## REFERENCES

[1] McQuiston, H., Plastics and the Mid-80's Automobile, *Plastics Engineering* (October 1979), pp. 22–29.

[2] Chasing After the Ideal Design, *Plastics Design Forum* (May/June 1980), pp 58–62

[3] Crate, J. H., Designing with Nylon, in *Nylon Plastics* (M. I. Kohan, ed.), Wiley, New York (1973), p. 589.

[4] Baer, E., *Engineering Design for Plastics*, Krieger Publishing Co., Melbourne (1975, reprint of 1964 edition).

[5] Beck, R. D., *Plastics Product Design*, 2nd ed., Van Nostrand Reinhold, New York (1980).

[6] Benjamin, B. S., *Structural Design with Plastics*, 2nd ed., Van Nostrand Reinhold, New York (1981).

# section vi

# appendices

# atoms commonly referred to in polymer work

## appendix a

| Atom | Symbol | Most Common Valence | Atomic Weight |
|------|--------|---------------------|---------------|
| Carbon | C | $-\overset{\vert}{\underset{\vert}{C}}-$ | 12.011 |
| Chlorine | Cl | Cl— | 35.453 |
| Fluorine | F | F— | 18.998 |
| Hydrogen | H | H— | 1.008 |
| Nitrogen | N | $\overset{\vert}{\underset{/\ \backslash}{N}}$ | 14.007 |
| Oxygen | O | —O— | 15.999 |
| Phosphorus | P | $-\overset{\vert}{\underset{\vert}{P}}=$ | 30.974 |
| Silicon | Si | $-\overset{\vert}{\underset{\vert}{Si}}-$ | 28.086 |
| Sulfur | S | —S— | 32.064 |
| | | $-\overset{\parallel}{\underset{\parallel}{S}}-$ | |

# organic chemistry nomenclature

## appendix b

I. *Hydrocarbons*

The simplest organic compounds contain only hydrogen and carbon, and these are called hydrocarbons. They include benzene:

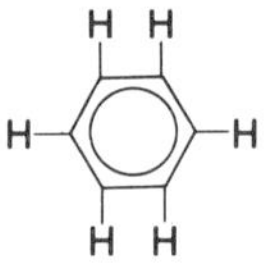

alkanes, alkenes, dienes, alkynes, cyclic aliphatics, and arenes. Benzene is an aromatic hydrocarbon. The others (with the exception of arenes) are aliphatic hydrocarbons. These various groups are described below.

(A) Alkanes

(1) Definition

The alkanes (paraffins) are aliphatic hydrocarbons that contain no doubly or triply bonded carbon atoms. They are said to be saturated and are represented by the general formula $C_nH_{2n+2}$.

(2) *Name* — *Structure*

1. Methane

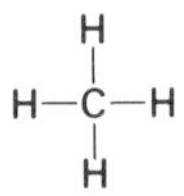

2. Ethane

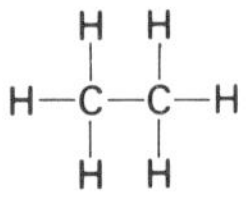

(2)    *Name*                                   *Structure*

3. Propane

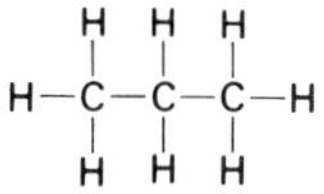

H—C—C—C—H with H atoms above and below each carbon:

```
     H   H   H
     |   |   |
H — C — C — C — H
     |   |   |
     H   H   H
```

4. Butane
There are two isomeric butanes:

   a. *n*-Butane, where "n" stands
   for normal (it is a straight
   chain)

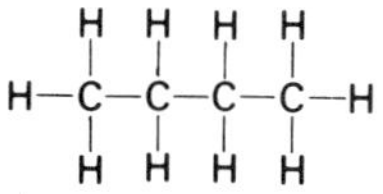

```
     H   H   H   H
     |   |   |   |
H — C — C — C — C — H
     |   |   |   |
     H   H   H   H
```

   b. Isobutane (it is a branched
   chain)

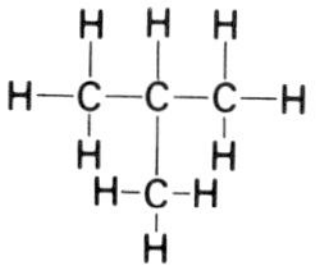

```
     H   H   H
     |   |   |
H — C — C — C — H
     |   |   |
     H  H-C-H H
         |
         H
```

5. Pentane
There are three isomeric pentanes:

   a. *n*-Pentane

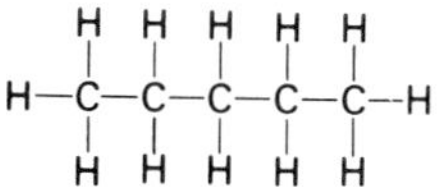

```
     H   H   H   H   H
     |   |   |   |   |
H — C — C — C — C — C — H
     |   |   |   |   |
     H   H   H   H   H
```

   b. Isopentane

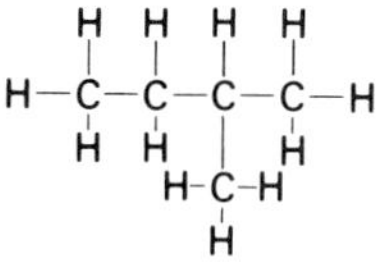

```
     H   H   H   H
     |   |   |   |
H — C — C — C — C — H
     |   |   |   |
     H   H  H-C-H H
             |
             H
```

   c. Neopentane

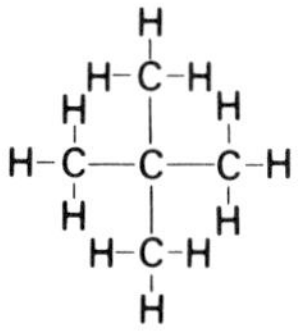

```
             H
             |
           H-C-H
         H   |   H
         |   |   |
   H - C — C — C - H
         |   |   |
         H H-C-H H
             |
             H
```

6. Hexane

```
     H   H   H   H   H   H
     |   |   |   |   |   |
H — C — C — C — C — C — C — H
     |   |   |   |   |   |
     H   H   H   H   H   H
```

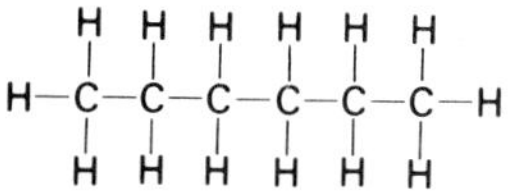

There are five isomeric hexanes.

7. Heptane

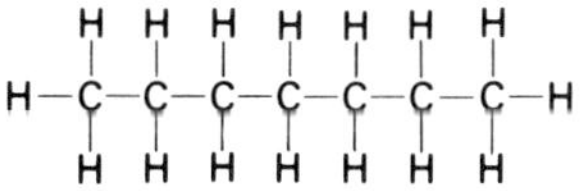

```
     H   H   H   H   H   H   H
     |   |   |   |   |   |   |
H — C — C — C — C — C — C — C — H
     |   |   |   |   |   |   |
     H   H   H   H   H   H   H
```

There are nine isomeric
heptanes.

(2)    *Name*                                              *Structure*

8.· Octane

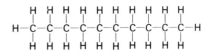

9. Nonane

10. Decane

There are 75 isomeric
decanes.

(B) Alkenes
  (1) Definition
      The alkenes are aliphatic hydrocarbons that contain a carbon–carbon
      double bond. These compounds are sometimes called olefins, and their
      general formula is $C_n H_{2n}$.
  (2)    *Name*                                              *Structure*

1. Ethylene

2. Propylene

3. Butene
   There are four isomeric butenes:

a. 1-Butene

b. *cis*-2-Butene

(*cis* = on one side of the
double bond)

(2)     *Name*                                      *Structure*

c. *trans*-2-Butene

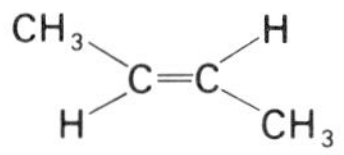

$$CH_3\diagdown_{\phantom{C}}\underset{H}{\overset{}{C}}=\underset{CH_3}{\overset{}{C}}\diagup^H$$

(*trans* = across the
double bond)

d. Isobutylene

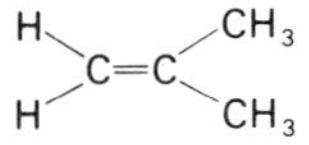

$$H\diagdown_{H}\overset{}{C}=\overset{}{C}\diagup^{CH_3}_{CH_3}$$

(C)  Dienes
   (1) Definition
      Dienes are aliphatic hydrocarbons that contain two carbon–carbon double
      bonds. There are three classes within this group, which are based on the
      arrangement of the double bonds. Conjugated double bonds are those that
      alternate with single bonds, as in

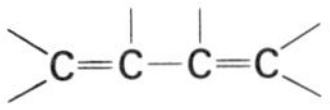

$$\diagup C=C-C=C\diagdown$$

      Isolated double bonds are those separated by more than one single bond,
      as in

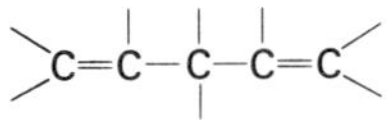

$$\diagup C=C-C-C=C\diagdown$$

      and cumulated double bonds are those without single bonds between them,
      as in

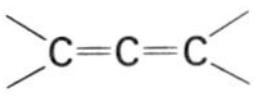

$$\diagup C=C=C\diagdown$$

      In polymer work the conjugated systems are the most common.
   (2)     *Name*                                  *Structure*

   1. 1,3-Butadiene

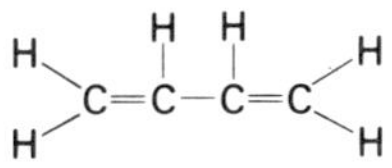

$$H\diagdown_{H}C=\overset{H}{\underset{}{C}}-\overset{H}{\underset{}{C}}=C\diagup^{H}_{H}$$

   2. 1,3-Isoprene

$$H\diagdown_{H}C=\overset{CH_3}{\underset{}{C}}-\overset{H}{\underset{}{C}}=C\diagup^{H}_{H}$$

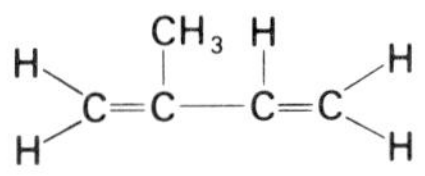

(D) Alkynes
  (1) Definition
      The alkynes are aliphatic hydrocarbons that contain a carbon–carbon
      triple bond. Their general formula is $C_n H_{2n-2}$.
  (2)        *Name*                              *Structure*

      1. Acetylene                          $H-C\equiv C-H$

      2. Propyne                            H–C≡C–C–H (with H above and below the terminal C)

      3. 1-Butyne                           H–C≡C–C–C–H (with H H above and H H below)

      4. 1-Pentyne                          H–C≡C–C–C–C–H (with H H H above and H H H below)

      5. 1-Hexyne                           H–C≡C–C–C–C–C–H (with H H H H above and H H H H below)

      6. 2-Butyne                           H–C–C≡C–C–H (with H above and below each terminal C)


(E) Cyclic Aliphatic Hydrocarbons
  (1) Definition
      Cyclic aliphatic hydrocarbons are ringed compounds whose names
      correspond to the open-chain hydrocarbons that have the same number of
      carbon atoms.
  (2)        *Name*                              *Structure*

      1. Cyclopropane                       $H_2C$ ... $CH_2$ ... $H_2C$ (triangular ring)

      2. Cyclobutane                        $H_2C-CH_2$ / $H_2C-CH_2$ (square ring)

      3. Cyclopentane                       $CH_2$ / $H_2C$ $CH_2$ / $H_2C-CH_2$ (pentagon ring)

      4. Cyclohexane                        $CH_2$ / $H_2C$ $CH_2$ / $H_2C$ $CH_2$ / $CH_2$ (hexagon ring)
         (*Note*: This is not a benzene ring.)

|     | (2) *Name* | *Structure* |

5. Cycloheptane

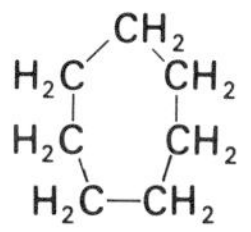

(F) Arenes
  (1) Definition
      Arenes have structures that are partially aliphatic and partially aromatic
      hydrocarbon.
  (2) *Name*                                              *Structure*

1. Toluene
   (methylbenzene)

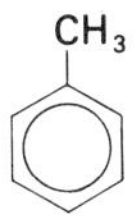

2. *o*-Xylene
   (1, 2-dimethylbenzene)
   ("*o*" = ortho)

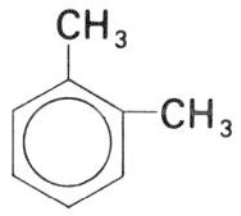

3. *m*-Xylene
   (1, 3-dimethylbenzene)
   ("*m*" = meta)

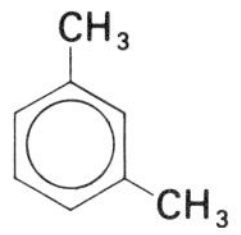

4. *p*-Xylene
   ("*p*" = para)

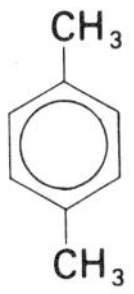

5. Ethylbenzene

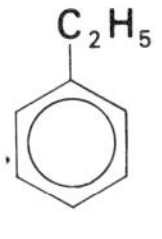

6. *n*-Propylbenzene

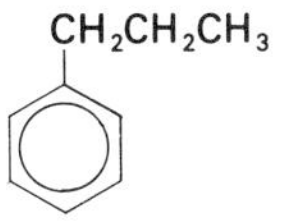

7. Styrene
   (vinylbenzene or
   phenylethylene)

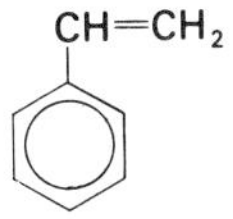

II. *Compounds Containing Oxygen or Nitrogen*
   In addition to the basic hydrocarbons just discussed, there are a variety of
   industrially significant organic compounds that contain oxygen and/ or nitrogen in

addition to carbon and hydrogen. These include the alcohols, ethers, epoxides, carboxylic acids, amines, aldehydes, ketones, glycols, phenols, nitrites, and peroxides. They are all described below.

(A) Alcohols
   (1) Definition

      Alcohols are compounds of the general formula ROH, where the portion of R attached directly to the hydroxyl (OH) is not a benzene ring. All alcohols contain the hydroxyl group, and it determines the properties characteristic of this family.

   (2)    *Name*                             *Structure*

       1. Methyl alcohol
          (methanol)

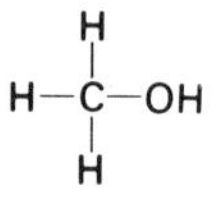

       2. Ethyl alcohol
          (ethanol)

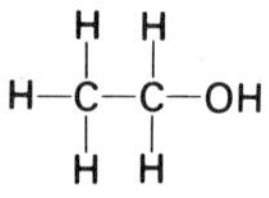

       3. Propyl alcohol
         There are two isomers:

          a. *n*-Propyl alcohol

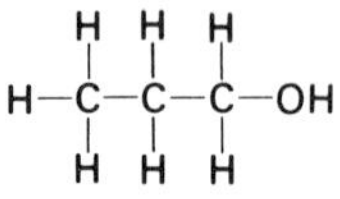

          b. Isopropyl alcohol

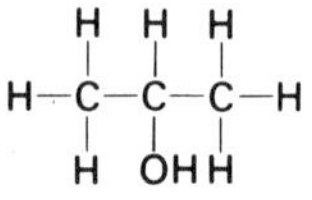

       4. Benzyl alcohol

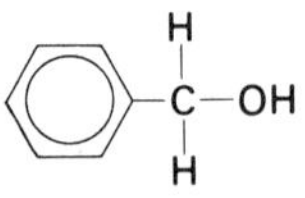

(B) Ethers
   (1) Definition

      Ethers have the general formula $\overset{R'}{\underset{R}{\diagdown}}$O, where R and R' may or may not be the same group.

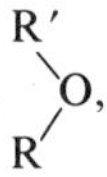

   (2)    *Name*                             *Structure*

       1. Methyl ether

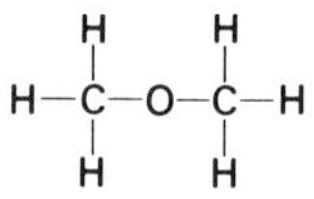

(2)    *Name*                                                    *Structure*

2. Ethyl ether

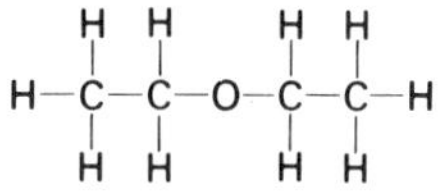

$$H-\underset{\underset{H}{|}}{\overset{\overset{H}{|}}{C}}-\underset{\underset{H}{|}}{\overset{\overset{H}{|}}{C}}-O-\underset{\underset{H}{|}}{\overset{\overset{H}{|}}{C}}-\underset{\underset{H}{|}}{\overset{\overset{H}{|}}{C}}-H$$

3. Phenyl ether

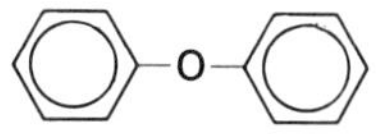

4. *n*-Propyl ether

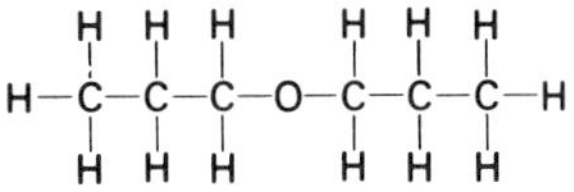

$$H-\underset{\underset{H}{|}}{\overset{\overset{H}{|}}{C}}-\underset{\underset{H}{|}}{\overset{\overset{H}{|}}{C}}-\underset{\underset{H}{|}}{\overset{\overset{H}{|}}{C}}-O-\underset{\underset{H}{|}}{\overset{\overset{H}{|}}{C}}-\underset{\underset{H}{|}}{\overset{\overset{H}{|}}{C}}-\underset{\underset{H}{|}}{\overset{\overset{H}{|}}{C}}-H$$

5. Vinyl ether

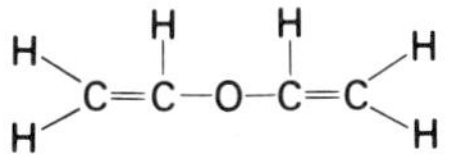

(C) Epoxides
  (1) Definition
      Epoxides are compounds that contain the oxirane ring:

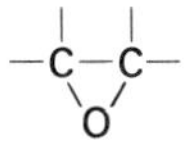

(2)    *Name*                                                    *Structure*

1. Ethylene oxide

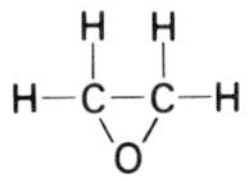

2. Styrene oxide

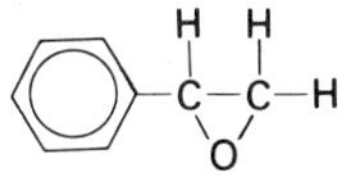

3. Diglycidyl ether of bisphenol A

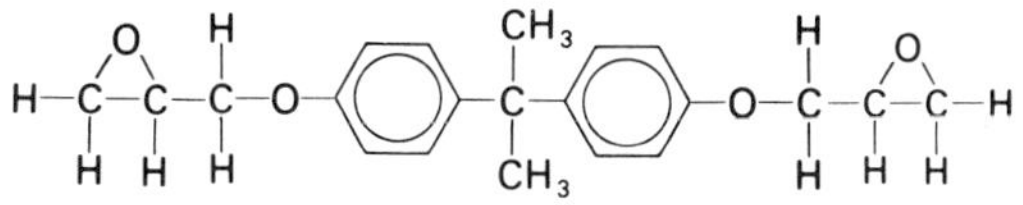

(D) Carboxylic Acids
  (1) Definition

Carboxylic acids have the $-\overset{\overset{\displaystyle O}{\|}}{C}-OH$ group on one or more ends of the molecule.

(2)     *Name*                                    *Structure*

1. Formic acid

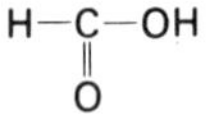

$$H-\underset{\underset{O}{\|}}{C}-OH$$

2. Acetic acid

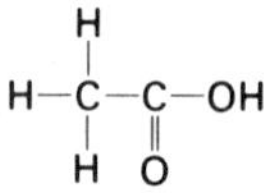

$$H-\underset{\underset{H}{|}}{\overset{\overset{H}{|}}{C}}-\underset{\underset{O}{\|}}{C}-OH$$

3. Propionic acid

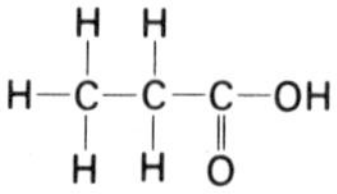

$$H-\underset{\underset{H}{|}}{\overset{\overset{H}{|}}{C}}-\underset{\underset{H}{|}}{\overset{\overset{H}{|}}{C}}-\underset{\underset{O}{\|}}{C}-OH$$

4. Butyric acid

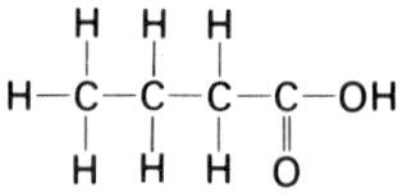

$$H-\underset{\underset{H}{|}}{\overset{\overset{H}{|}}{C}}-\underset{\underset{H}{|}}{\overset{\overset{H}{|}}{C}}-\underset{\underset{H}{|}}{\overset{\overset{H}{|}}{C}}-\underset{\underset{O}{\|}}{C}-OH$$

5. Adipic acid

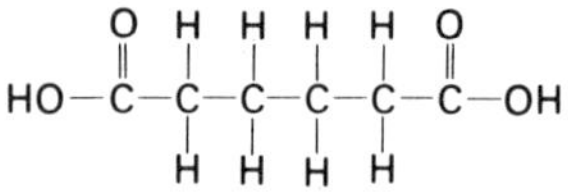

$$HO-\underset{\underset{O}{\|}}{C}-\underset{\underset{H}{|}}{\overset{\overset{H}{|}}{C}}-\underset{\underset{H}{|}}{\overset{\overset{H}{|}}{C}}-\underset{\underset{H}{|}}{\overset{\overset{H}{|}}{C}}-\underset{\underset{H}{|}}{\overset{\overset{H}{|}}{C}}-\underset{\underset{O}{\|}}{C}-OH$$

6. Sebacic acid

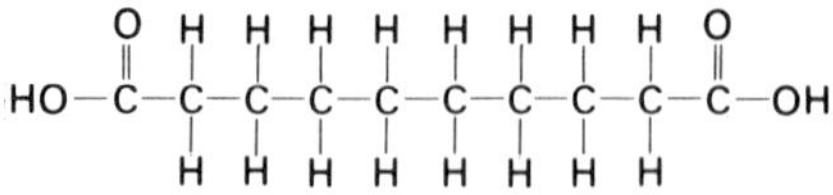

$$HO-\underset{\underset{O}{\|}}{C}-C-C-C-C-C-C-C-C-OH$$
(with H substituents)

7. Stearic acid

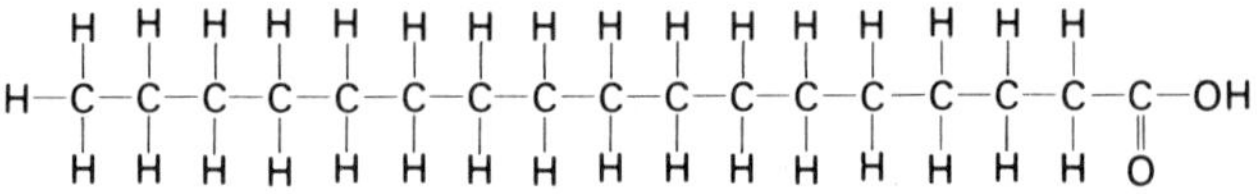

$$H-C-C-C-C-C-C-C-C-C-C-C-C-C-C-C-C-C-OH$$
(long chain with H substituents)

8. Benzoic acid

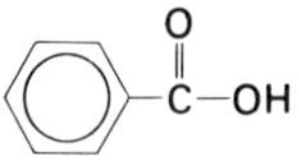

(benzene ring)$-\underset{\underset{}{}}{\overset{\overset{O}{\|}}{C}}-OH$

9. *o*-Toluic

$H-\overset{\overset{H}{|}}{\underset{\underset{H}{|}}{C}}-H$ on benzene ring with $-\underset{\underset{O}{\|}}{C}-OH$

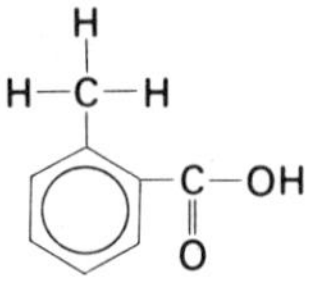

10. Terephthalic acid

benzene ring with $\overset{OH}{\underset{}{C=O}}$ and $\underset{OH}{\overset{}{C=O}}$

(E) Amines
  (1) Definition
    The substitution of organic groups for hydrogen atoms in ammonia ($NH_3$)
    produces the amines.
  (2)    *Name*                              *Structure*

1. Methylamine

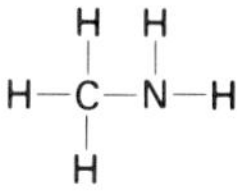

2. Dimethylamine

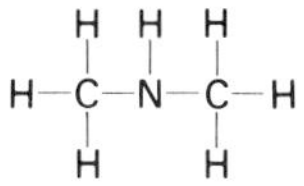

3. Trimethylamine

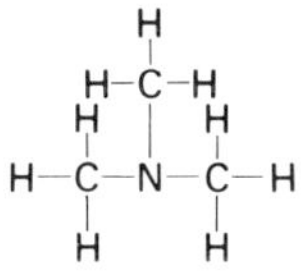

4. Ethylenediamine

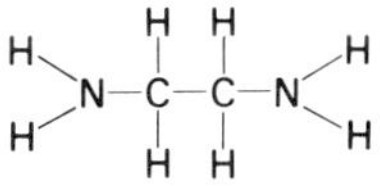

5. Diethylenetriamine
   (DTA)

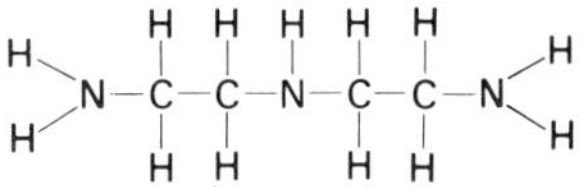

6. Hexamethylenediamine

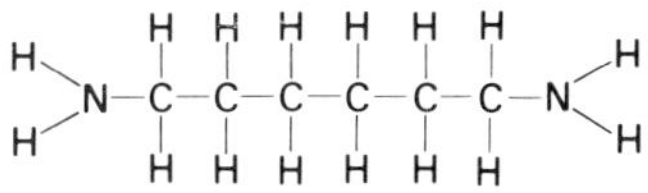

7. *m*-Phenylenediamine

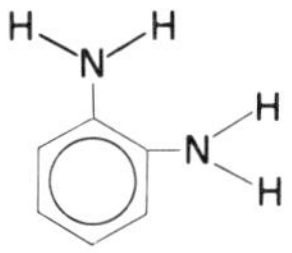

8. Urea

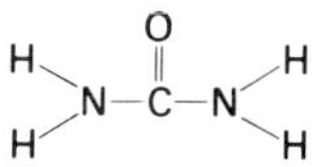

9. Melamine

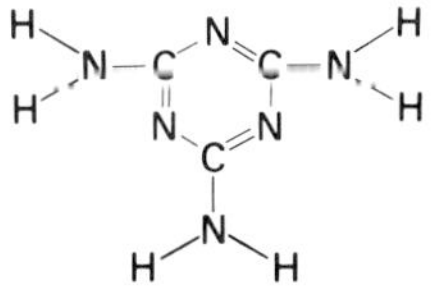

(F) Aldehydes
    (1) Definition

Aldehydes have the general formula $R-\overset{\displaystyle O}{\overset{\|}{C}}-H$, where R can be hydrogen or an organic group. The double bonded carbon-oxygen (C=O) is the carbonyl group.

    (2)   *Name*                                               *Structure*

       1. Formaldehyde

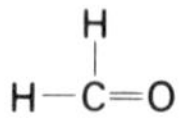

       2. Acetaldehyde

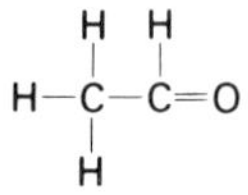

       3. Propionaldehyde

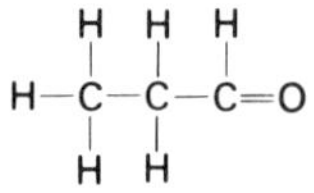

       4. *n*-Butyraldehyde

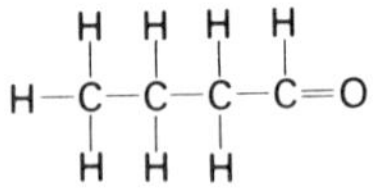

(G) Ketones
    (1) Definition

Ketones have the general formula $R-\overset{\displaystyle O}{\overset{\|}{C}}-R'$, where R and R' are organic groups, but not hydrogen.

    (2)   *Name*                                               *Structure*

       1. Acetone

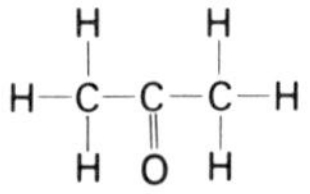

       2. Methyl ethyl ketone
          (MEK)

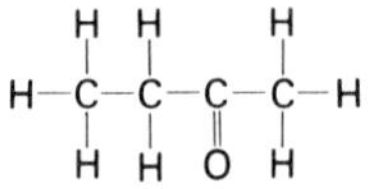

       3. Methyl *n*-propyl
          ketone

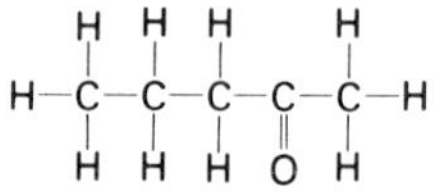

(H) Glycols
    (1) Definition
    Glycols are alcohols containing two or more hydroxyl (OH) groups.

(2)    *Name*                                                      *Structure*

1. Ethylene glycol

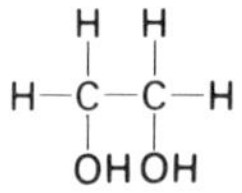

$$\begin{array}{c} \text{H}\quad\text{H} \\ |\quad\ | \\ \text{H}-\text{C}-\text{C}-\text{H} \\ |\quad\ | \\ \text{OH}\,\text{OH} \end{array}$$

2. Propylene glycol
   (1,2-propanediol)

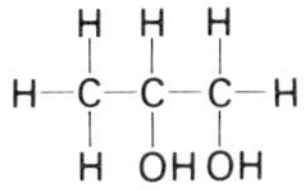

$$\begin{array}{c} \text{H}\ \ \text{H}\ \ \text{H} \\ |\quad|\quad| \\ \text{H}-\text{C}-\text{C}-\text{C}-\text{H} \\ |\quad|\quad| \\ \text{H}\ \ \text{OH}\,\text{OH} \end{array}$$

3. Glycerol
   (1,2,3-propantriol)

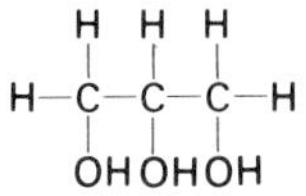

$$\begin{array}{c} \text{H}\ \ \text{H}\ \ \text{H} \\ |\quad|\quad| \\ \text{H}-\text{C}-\text{C}-\text{C}-\text{H} \\ |\quad|\quad| \\ \text{OH}\,\text{OH}\,\text{OH} \end{array}$$

(I) Phenols
   (1) Definition
      Phenols have the general formula ROH, where R is generally either a
      benzene ring or a benzene ring with organic or inorganic substituents. The
      benzene ring must be attached directly to the hydroxyl group. Note that
      these are different from alcohols.

   (2)    *Name*                                                   *Structure*

1. Phenol

2. *m*-Cresol
   (methylphenol)

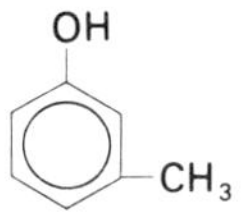

3. Resorcinol

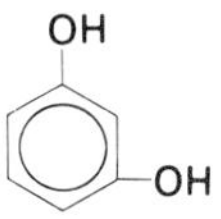

4. Hydroquinone

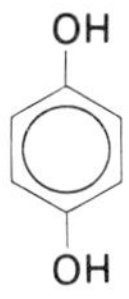

5. Bisphenol A

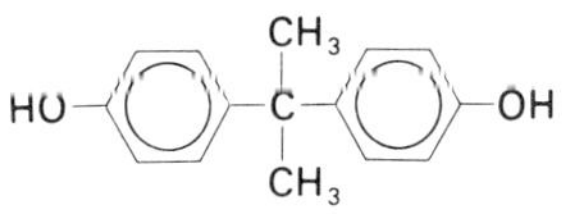

(J) Nitriles
    (1) Definition
        Nitriles are compounds that contain the $-C{\equiv}N$ group.
    (2)    *Name*                          *Structure*

        1. Acrylonitrile

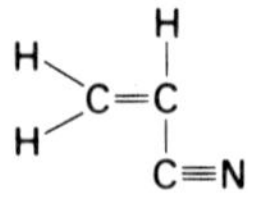

        2. Azobisisobutyronitrile

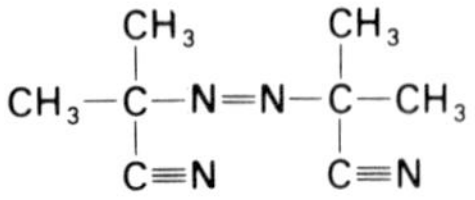

(K) Peroxides
    (1) Definition
        Peroxides are compounds containing the $-O{-}O-$ linkage. They are generally very reactive.
    (2)    *Name*                          *Structure*

        1. Benzoyl peroxide

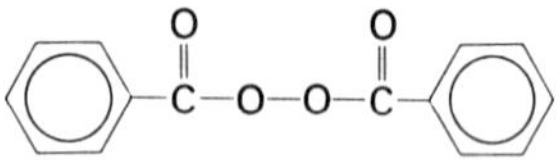

III. *Miscellaneous Compounds*
    The following are compounds that do not readily fall into any of the previous categories but are referred to in this text.
    (A) Halogenated Compounds
        (1) Definition
            These are organic compounds containing fluorine, chlorine, or bromine.
        (2)    *Name*                          *Structure*

        1. Vinyl chloride

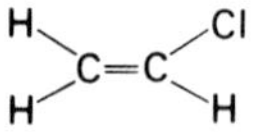

        2. Vinylidene chloride

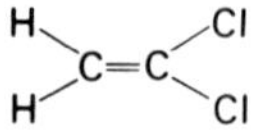

        3. Phosgene

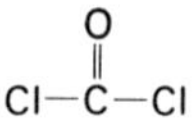

        4. Chloroprene

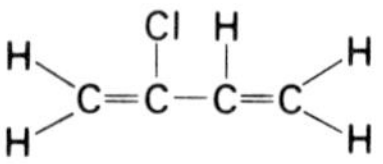

        5. Tetrafluoroethylene

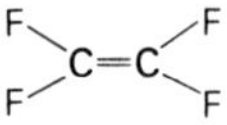

(B) Other Compounds Not Listed Elsewhere
   (1)    *Name*                                    *Structure*

1. Methyl methacrylate

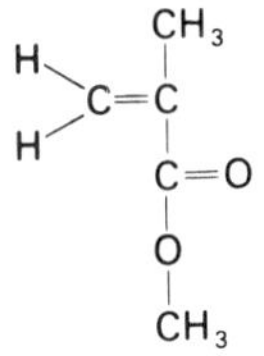

2. Tricresyl phosphate
   (a mixture of ortho, meta,
   and para isomers)

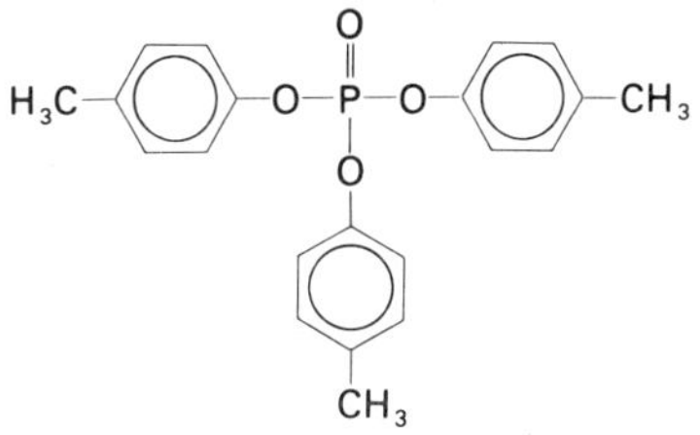

# si conversions

# appendix c

The following chart contains the conversion factors necessary to transform the SI units in this text into other selected units. This material was adapted from ASTM E 380-79, Standard for Metric Practice [1].

| Quantity or Property | To Convert from | To | Multiply by |
|---|---|---|---|
| Bond angle | rad | degrees | $5.730 \times 10^{1}$ |
| Bond length | m | Å | $1.000 \times 10^{10}$ |
| Bond strength | kJ/mol | kcal/mol | $2.390 \times 10^{-1}$ |
| Density | $kg/m^3$ | $g/cm^3$ | $1.000 \times 10^{-3}$ |
| | $kg/m^3$ | $lb/ft^3$ | $6.243 \times 10^{-2}$ |
| Dielectric strength | V/m | V/mil | 25.4 |
| Force | N | dyne | $1.000 \times 10^{5}$ |
| | N | lbf | $2.248 \times 10^{-1}$ |
| | N | kip | $2.248 \times 10^{-4}$ |
| Heat of reaction | J/kg | cal/g | $2.390 \times 10^{-4}$ |
| | J/kg | BTU/lb | $4.302 \times 10^{-4}$ |
| Modulus, strength | Pa | psi | $1.450 \times 10^{-4}$ |
| Specific heat | J/kg·K) | BTU/(lb·°F) | $2.390 \times 10^{-4}$ |
| | J/kg·K) | cal/(g·°C) | $2.390 \times 10^{-4}$ |
| Temperature | K | °C | subtract 273.15 |
| | K | °F | multiply by 1.8 and then subtract 459.67 |
| Thermal conductivity | W/(m·K) | (BTU·ft)/(h·$ft^2$·°F) | $5.782 \times 10^{-1}$ |
| | W/(m·K) | cal/(cm·s·°C) | $2.390 \times 10^{-3}$ |

| Quantity or Property | To Convert from | To | Multiply by |
|---|---|---|---|
| Thermal expansion | (m/m)/K | (in/in)/°F | $5.556 \times 10^{-1}$ |
|  | (m/m)/K | (m/m)/°C | 1.000 |
| Volume resistivity | $\Omega \cdot m$ | $\Omega \cdot cm$ | $1.000 \times 10^{2}$ |
|  | $\Omega \cdot m$ | $(\Omega \cdot cir\ mil)/ft$ | $6.015 \times 10^{8}$ |

## REFERENCES

[1] American Society for Testing and Materials, *ASTM Standards, Part 41*, Philadelphia (1981).

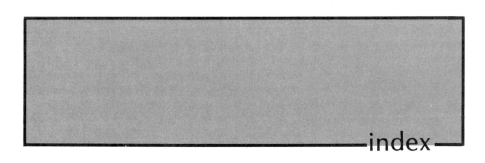

index